WATER-ROCK INTERACTIONS

A case study in a very
low grade metamorphic
shale catchment in the
Ardennes, NW Luxembourg

J. M. Verstraten

Universiteit van Amsterdam
Fysisch Geografisch en
    Bodemkundig Laboratorium,
Dapperstraat 115
Amsterdam
Holland

GEO BOOKS
Geo Abstracts Ltd
Norwich

Published by Geo Abstracts, Norwich.

British Geomorphological Research Group, Research Monograph
Series 2.

Published in two editions.
This edition hardback complete with appendix (pp 177-244)
ISBN 0 86094 042 X

Paperback with Appendix on microfiche
ISBN 0 84094 043 8

Also in this series:
No. 1. K. Gilman & M.D. Newson, Soil pipes and pipeflow.
       A hydrological study in upland Wales. 1980.

Other titles in preparation

Geo Abstracts, University of East Anglia, Norwich, NR4 7TJ,
England.

80 008365

ACKNOWLEDGEMENTS

I am greatly indebted to the late Dr. Ir. J. van Schuylenborgh, who introduced me into the field of aquatic chemistry in relation to weathering and soil genesis. Field and laboratory work for this thesis was carried out under his supervision.

I wish to record my sincere appreciation to Prof. Dr. P. D. Jungerius for his critical reading of the manuscript, and for his advice in problems concerning landscape development and actuo geomorphology.

Special thanks are due to Prof. W. van Tongeren for his critical remarks and helping with the correction of the text, and to Prof. Dr. Ir. A. P. A. Vink for reading the manuscript and providing many helpful comments.

Also special thanks are due to Dr. A. C. Imeson for his assistance in the field, encouragement, many fruitful discussions on problems concerning the hydrology and the improvement of the text.

Thanks are due to:

Mr. F. J. P. M. Kwaad for the discussion on landscape history
Mr. E. A. Kummer and Mr. B. de Leeuw for the X-ray diffraction analysis
Mr. H. J. Mücher for his assistance and valuable comments on the micromorphological study
Dr. J. Sevink for the stimulating discussions on soil classification and genesis
Dr. R. T. Slotboom for counting and interpretation of the pollen samples
Mr. R. van der Tuin and Mrs. M. J. van Wijk-Groot for the assistance concerning computerised data processing
Dr. J. Verhofstad for reading parts of the manuscript
Mr. H. W. Flohr, Mr. L. Hoitinga, Mr. L. de Lange, Mr. P. Wartenbergh, Mr. N. de Wilde de Ligny & Mr. A. J. van Wijk, whose laboratory work formed the foundation of this study
Mr. C. Zeegers for preparing thin sections
Mr. A. J. van Geel and Mr. C. J. Snabilie for drawing and assistance in printing
Mr. A. Eikeboom for the photography
Mrs. M. C. G. Keijzer-van der Lubbe for typing and correcting the manuscript in such a short period
Mr. L. Huysen and Mr. A. P. J. van Wanrooy for handling the offset procedures

Finally I wish to express my deep gratitude to my wife Marianne whose support contributed much towards the completion of this study.

EDITOR'S PREFACE.

This monograph series is organised by the British
Geomorphological Research Group and its purpose is twofold.
Firstly, the monographs may report on a substantial original research endeavour such as might be undertaken on one
particular landform type, area or other topic. Secondly,
they may review the current state of knowledge in a field
of research. In both cases the series provides an outlet
for geomorphological publication which is longer than is
normally considered for papers in journals. In the case
of research reports the authors are able to report in depth
and in detail on a substantial research project or group
of allied projects. In the second case the monographs provide perspectives on the state of current knowledge in particular fields of geomorphology, highlighting recent prog-
ress and future research needs. The series covers the field
of international geomorphology in its widest sense, including process, form, technique, theory and applied geomorphology.

BGRG Research Monograph Editor
Dr. Stephen T. Trudgill
Department of Geography
University of Sheffield, S10 2TN, U.K.

PART A

1. INTRODUCTION

1. General

The present study deals mainly with the water-rock interactions in (very) low-grade metamorphic shales and the resultant soils in a small, 16.9 ha, completely forested catchment in the Luxembourg Ardennes (fig. 1). The forested Haarts catchment was selected in order to exclude recent anthorpogenic influences by agricultural practices, and to obtain a good insight into a natural ecosystem.

Until now work on soil genesis has mostly been carried out on solid phases of soils and parent materials. Little attention was paid to the weathering agent, the liquid phase, with its solutes. In this study the emphasis was placed on the study of the spring- and riverwater chemistry, as in the Haarts catchment these waters represent the weathering environment and give information on the actual weathering processes. The study of the water chemistry, in addition to that of the weathering materials, has many advantages. In particular, for relative young soils like those in the Haarts catchment, due to the large mass ratio of solids versus solution, undetectable small changes in the composition of the rock and soils show up as easily detectable, relatively large changes in the composition of the soil-, spring- and riverwaters. This approach requires water sampling at specific intervals throughout the year and continuous measurements of discharge at the catchment outlet. With this procedure chemical weathering processes in the soil environment under various hydrological conditions can be traced as well as the output of solutes from the system. It also enables the calculation of a geochemical balance for the forested catchment if additional information on the weathering products exists.

Geomorphological investigations in the Birbaach catchment carried out by the Laboratory for Physical Geography and Soil Science enabled the reconstruction of the complex Late Quaternary landscape history of the catchment. It appeared that the changing environmental conditions during this period resulted in erosion and deposition of the weathering materials; they had great influence on the composition of the soils. Thus only a multidisciplinary, physical geographical approach will lead to a more complete picture of the weathering system. Consequently, all of the relevant landscape forming factors operating during the Late Quaternary, and their influences on the saprolites were studied and will be discussed in this thesis.

2. Location of the study area

The location of the area studied is indicated in fig. 1. The forested and unhabitated Haarts catchment lies at 49⁰56' N.L. and 5⁰52' E.L. and belongs to the southern part of the Oesling, the Luxembourg Ardennes. Morphologically, it is part of an undulating plateau landscape with steeply incised river valleys.

1

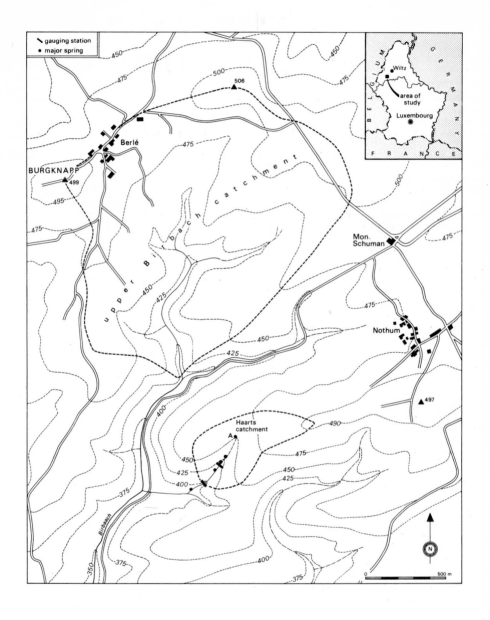

Figure 1   Location of the area studied (Haarts catchment
with major springs and gauging stations)

2

11. FIELD AREA

1. CLIMATE

1.1 General

Luxembourg has a humid temperate climate, which can be classified according to Köppen's system as Cfb. The Oesling is a part of the "Region Ardennaise", one of the three climatic regions in Luxembourg. This region is predominantly cool and hazy, characterized by SW and W winds and strong changes in temperature and humidity. The insolation is very variable and rain and snow are frequent and abundant (Lahr 1964).

Long term meteorological records are available from Berlé (49°.57' N and 05°51'E; altitude 495 m), an important meteorological station in the Luxembourg network located only a few hundred meters NW of the study area. The highest parts of the Haarts catchment are at the same altitude as Berlé, which is, excluding the effect of vegetation probably representative of the area.

The average annual precipitation for the years 1931-1960 was 880 mm. Although this is fairly uniformly distributed throughout the year, maxima tend to occur in July-August (local storms) and November-December (frontal) (table 1). The number of days with rain or snow is at maximum in November-December, but otherwise these are uniformly distributed throughout the year (table 1). A large part of the rain during the summer falls in relatively short showers, which is demonstrated by the maximum precipitation per day data, also shown in table 1. However, a rainfall intensity of more than 60 mm per hour is rare in Luxembourg (Lahr 1964).

The mean monthly relative humidity is continuously high (70-90%) with a yearly average of 80 percent. The mean cloudiness also shows a maximum in November-December, but for the rest of the year it is rather constant. The wind at Berle is predominantly from the south west (table 2). The average annual temperature is 8.0 C but the mean monthly temperature shows rather large fluctuations (-0.7° - 16.3°C). The influence of the altitude is also obviously demonstrated by the great differences between the absolute maximum and minimum temperatures (table 1). Some supplementary temperature data are given in table 1.

1.2 Climatic criteria for the classification of the soils

Unfortunately no data on the soil climate are known from the Luxembourg Ardennes. However, some comments will be made with respect to the climatic criteria used for the classification of the soils at suborder and great group level in the Soil Taxonomy (1975).

From data on soil temperature in Western Germany (Dov Ashbel *et al* 1965) and the non-coastal part of Belgium (Deckers 1966) compared with the mean monthly air temperatures it can be concluded that the mean annual soil temperature is somewhat higher than 8°C but lower than

3

Table 1 Climatological conditions at Berle*

	period	J	F	M	A	M	J	J	A	S	O	N	D	Year
mean monthly precipitation; mm	1931-60	73	76	51	54	70	74	89	95	71	60	79	90	880
mean number of days with rain or snow	1931-60	21	19	16	14	16	14	15	17	16	17	21	22	208
maximum precipitation per day; mm	1931-68	32	46	30	32	45	60	72	52	35	36	54	32	72
cloudiness, in 10ths of the cloud-covered sky	1949-60	7.9	7.6	6.4	6.2	6.6	6.7	6.5	6.4	6.3	6.7	8.4	8.7	7.0
mean monthly temperature; °C	1931-60	-0.7	0.2	3.8	7.3	11.7	14.5	16.3	15.9	13.3	8.5	4.1	0.4	8.0
maxima of "absolute" temperature; °C	1949-68	13.4	15.5	20.5	27.9	27.9	31.1	33.3	30.6	29.1	22.3	17.4	12.2	33.3
minima of "absolute" temperature; °C	1949-68	-17.1	-21.2	-12.6	-5.7	-2.7	-0.7	4.0	4.2	0.0	-5.7	-10.8	-14.0	21.2
mean number of days with max.temp. ≥ 25°C	1949-68	0	0	0	<<1	1	3	5	3	1	0	0	0	13
mean number of days with min.temp. < 0°C	1949-68	23	18	14	7	1	<<1	0	0	<<1	2	12	20	97
mean number of days with mean temp. ≥ 10°C	1949-68	0	<<1	2	7	18	27	30	31	25	12	1	<<1	153

*From Lahr (1964) and Faber (1971).

Table 2   Frequency of the wind directions in percentages
           at Berlé*(1949-68)

N	NE	E	SE	S	SW	W	NW
10.3	11.1	10.5	7.9	14.2	22.6	14.2	9.2

* from Faber (1971)

15^0C and the difference between mean summer and mean winter
soil temperature is more than 5^0C at a depth of 50 cm or
at a lithic or paralithic contact, whichever is shallower.
So the soil temperature regime can be classified according
to the Soil Taxonomy (1975) as mesic. This conclusion
is corroborated by the temperatures of the spring waters in
the Haarts catchment, which fluctuates between 6.5^0C and
11^0C with an average of 8.5^0C.

       The soil moisture regime is classified as udic
(fig. 2). This moisture regime implies that in moist years
the soil moisture control section is not dry in any part
for as long as 90 days (cumulative) and because the soil
has a mesic temperature regime the soil moisture control
section is not dry in all parts for as long as 45
consecutive days in 4 months that follow the summer solstice
in 6 or more years out of 10. The udic moisture regime
requires, except for short periods, a three phase system,
solid-liquid-gas, in part but not necessarily in all of the
soil when the soil temperature is above $5°C$. The udic
moisture regime is common to the soils of the humid
climates that have well-distributed rainfall, or that have
sufficient rain in summer to ensure that the amount of
stored moisture plus rainfall is approximately equal to,
or exceeds, the amount of evapotranspiration (Soil
Taxonomy 1975). The Luxembourg Ardennes fulfil to these
requirements, which will be demonstrated in the next paragraph.

1.3 Climatic data for the years 1973-1975

       In order to give an indication of the climatic
conditions during the period of investigation, some climatic
parameters are presented in table 3 A-J. In general no
large deviations from the long term mean were measured as
far as the temperature is concerned (table 1). Only
January and February in 1974 and 1975 were warmer than
average and the summer of 1974 was relatively cool. From
data in table 3H it can be concluded that the years 1973
and 1975 were drier than the mean. For 1973 especially, the
winter, early spring and summer months were extremely dry.
In 1975 February, May and the summer months all received
less precipitation than the average. The year 1974 was
more normal as far as the average annual precipitation
is concerned, but in this year also the first five months
were relatively dry, while the second half of September
and October was very wet.

Table 3 Climatic data from Berlé 1973 to 1975

A. Mean monthly temperatures ($^{\circ}$C)

	J	F	M	A	M	J	J	A	S	O	N	D	Year
1973	-0.1	-0.3	3.1	4.1	11.1	15.1	15.3	17.6	13.9	7.3	2.7	-0.4	7.5
1974	3.0	2.2	4.7	7.8	9.8	13.2	13.7	15.5	11.2	4.0	3.9	4.0	7.8
1975	3.6	2.1	2.1	5.8	9.9	13.4	16.8	18.2	13.9	6.4	3.1	-0.9	7.9

B. Mean monthly maximum and minimum temperatures ($^{\circ}$C)

	J	F	M	A	M	J	J	A	S	O	N	D	Year
1973	2.2	2.1	7.7	7.8	15.3	19.7	19.8	23.3	19.0	11.1	5.4	1.7	11.3
	-1.9	-2.4	-0.4	1.0	6.8	10.2	11.6	12.9	9.6	4.1	0.1	-2.4	4.1
1974	4.8	4.8	5.0	12.8	14.2	17.9	17.9	20.5	15.4	6.2	6.3	5.8	11.0
	1.1	-0.1	1.7	3.5	5.5	8.4	9.8	11.6	7.9	2.2	1.5	2.2	4.6
1975	5.4	6.3	4.5	9.3	14.6	18.0	21.8	23.4	18.2	9.6	5.8	1.4	11.5
	1.5	-1.1	0.1	2.7	5.6	9.2	12.0	13.9	10.1	3.7	-0.6	-2.2	4.6

C. "Absolute" maximum and minimum temperatures ($^{\circ}$C)

	J	F	M	A	M	J	J	A	S	O	N	D	Year
1973	9.0	7.3	17.0	16.5	20.5	26.5	28.7	28.2	27.5	21.8	13.3	7.6	28.7
	-6.2	-8.3	-4.2	-3.7	0.8	5.0	8.0	7.0	3.9	-2.1	-7.5	-14.0	-8.3
1974	10.0	12.0	17.8	19.8	22.0	23.8	23.5	30.0	23.6	9.4	11.5	10.6	23.8
	-4.4	-5.4	-8.8	-2.0	0.0	2.5	6.0	4.5	1.8	-1.1	-5.6	-3.0	-8.8
1975	11.4	10.8	10.5	20.1	21.5	25.4	28.7	30.4	24.7	14.5	11.6	6.8	30.4
	-2.4	-5.2	-5.2	-4.6	-0.1	0.1	4.8	7.5	5.0	-2.5	-7.0	-9.0	-9.0

D. Mean monthly relative humidity

	J	F	M	A	M	J	J	A	S	O	N	D	Year
1973	94	92	79	83	76	72	77	67	76	84	89	91	82
1974	95	87	88	71	74	76	79	76	83	92	91	92	84
1975	90	73	88	80	73	76	72	71	80	87	90	93	81

E. Number of days that the soil is snow-covered

	J	F	M	A	M	J	J	A	S	O	N	D	Year
1973	7	12	0	0	0	0	0	0	0	0	5	14	38
1974	0	2	5	0	0	0	0	0	0	1	1	3	12
1975	1	1	15	6	0	0	0	0	0	0	0	0	21

6

Table 3   (continued)

F. Number of days with rain or snow

	J	F	M	A	M	J	J	A	S	O	N	D	Year
1973	13	18	8	19	18	8	14	8	10	13	15	23	167
1974	24	13	15	8	14	15	17	10	21	29	23	26	215
1975	24	8	26	17	11	10	9	9	14	9	17	8	162

G. Maximum daily precipitation

	J	F	M	A	M	J	J	A	S	O	N	D	Year
1973	6.8	19.7	2.2	10.7	23.7	11.3	11.0	8.8	16.6	36.6	11.2	13.3	
1974	14.6	15.0	10.4	13.5	13.8	21.2	9.0	14.3	19.2	18.0	15.6	18.6	21.2
1975	20.2	7.2	24.7	13.4	15.6	27.7	10.9	23.2	21.9	4.3	21.2	7.5	23.2

H. Mean monthly precipitation (P) in mm

	J	F	M	A	M	J	J	A	S	O	N	D	Year
1973	29.0	49.0	6.1	45.7	84.2	34.5	60.7	22.0	40.8	80.6	67.8	85.7	606.1
1974	59.9	49.0	54.6	21.9	40.5	74.1	63.6	42.8	128.3	121.9	89.8	98.0	844.4
1975	82.9	18.5	84.6	54.1	38.1	65.7	51.4	51.0	89.2	14.7	88.1	16.8	655.1

I. Potential evapotranspiration (PE) in mm

	J	F	M	A	M	J	J	A	S	O	N	D	Year
1973	4.5	4.6	23.4	28.8	56.3	78.7	71.3	104.6	71.2	22.5	9.1	3.6	478.6
1974	3.6	9.0	18.5	54.9	60.0	65.0	51.7	70.1	38.7	7.7	6.1	5.1	390.4
1975	7.0	17.8	12.9	36.4	58.8	60.7	81.7	95.6	42.0	14.1	6.6	3.9	437.5

J. P-PE (mm)

	J	F	M	A	M	J	J	A	S	O	N	D	Year
1973	25	44	-17	17	28	-44	-11	-83	-30	58	59	82	128
1974	56	40	36	-33	-20	-9	12	-27	90	114	84	93	454
1975	76	1	72	38	-21	5	-30	-45	47	1	82	13	198

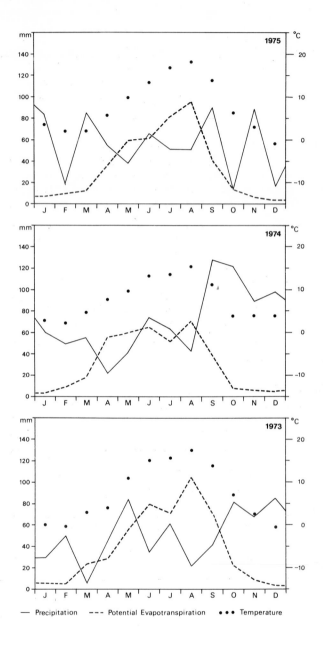

Figure 2   Mean monthly climatic data for the years 1973-1975

8

Despite the long history of rainfall measurement many problems and uncertainties about collection and accuracy of rainfall data still remain (Ward 1975). Although many factors like wind, type of gauge and its surrounding etc. influence the amount of precipitation measured, the level above ground at which the gauge is installed is one of the main factors. Also the influence of the relief on the precipitation has to be considered. Therefore a pluviograph (1.5 m above the ground level) and a rain gauge (40 cm above the ground level) were exposed at the lowest point of the catchment and another set was placed about 40 meters above this point. As the differences in amount of rain between the pluviograph and the rain gauge at one site were greater than the differences between the various sites, including the meteorological station at Berlé (90 m above the lowest point of the drainage basin) and no variations could be detected, the station at Berle was taken as representative for the whole catchment.

In order to establish the water balance and the hydrology in the Haarts catchment, the evapotranspiration has to be considered. As no measurements are available of the various factors which have to be used to calculate the potential evapotranspiration with a more physically sound method, for example Penman (Penman 1956), the empirical method of Haude (1954, 1963) has been used to obtain information on the potential evapotranspiration (table 3I). This method has given very good results in the Middle European climatic zone and has been successfully used in Western Germany (Dammann 1965). Uhlig (1954) found that Haude's method gives similar results as other potential evapotranspiration methods. This was also established by Van Zon (1975) for eastern Luxembourg.

The advantage of Haude's method is that for rather short time intervals, like one day, the potential evapo-transpiration can be calculated.

The formula $PE = f.d_{14}$ gives a daily potential evapo-transpiration where d_{14} is the daily saturation deficit at 14.00 h., which can be calculated from the relative humidity and the air temperature at that time. The daily saturation deficit has been calculated from

$RH = 100 \dfrac{e}{e_s}$ %, where e is the vapour pressure at a certain temperature, e_s is saturation vapour pressure at the same temperature

and $d = e_s - e$

The magnitude of the factor f is varied according to the season as shown in table 4.

Two opposing views prevail today about the interaction between potential evapotranspiration and soil moisture. The first concept is that of Viehmeyer (1927) and of

9

Table 4   Proportional factor of f

	Oct-Febry	March	April-May	June	July	August	Sept
f	0.26	0.33	0.39	0.37	0.35	0.33	0.31

Viehmeyer & Hendrikson (1927, 1955), who concluded from
"pot" experiments that the water loss was the same when the
soil moisture content had been reduced almost to the wilting
point as when the soil was at field capacity. Thus,
according to these authors, evapotranspiration takes place
at the potential rate throughout the period of soil
moisture depletion between the field capacity and the
wilting point. Gardner & Ehlig (1963) and Linsley et al
(1958) have shown that there is little variation in transpiration as the soil moisture content decreases until the
wilting point has been reached; thereafter, there is a
linear relationship between water content and transpiration
rate.

The second view is that evapotranspiration decreases
as the soil moisture decreases and has been discussed by
several authors (Kramer 1952; Thornthwaite 1954;
Bierhuizen et al 1960). Makkink & Van Heemst (1956) found
that the actual transpiration from grass-covered lysimeters
fell below potential evaporation as soon as the soil
moisture content decreases below field capacity. The rate
of this reduction was dependent on the moisture tension of
the soil and the rate of potential evapotranspiration.
Visser (1963, 1964) suggested that the shape of the diurnal
curve of the potential evapotranspiration is more important
than the daily totals of potential evapotranspiration. A
relation has been developed by Wartena & Veldman (1961)
between the amount of available water in the rootzone and
the actual evapotranspiration. They found that a decrease
of the percentage (V) of the total available soil moisture
to 70 per cent gives no reduction of the actual evapotranspiration. If V is equal or below 70 % the actual
evapotranspiration decreases linearly to one tenth of the
potential evapotranspiration at the wilting point.

The second view cannot be supported by physical theory,
whereas the Viehmeyer hypothesis can, especially where
roots ramify thoroughly throughout the soil (Penman 1969).
However, Viehmeyer's hypothesis frequently breaks down under
field conditions. Therefore Penman formulated the "root
constant" concept (Penman 1949) which proposes that because
moisture moves relatively slowly through the soil, the
readily available soil moisture is effectively restricted
to water in the immediate proximity of the root system,
and it will, therefore, be limited in amount partly by
depth of the soil available, and partly by the soil type.
Although the root constant is primarily a plant
characteristic it is modified by other factors such as soil
type and depth. The main factor affecting the root constant

is the depth of rooting and in the literature 25-30 cm
or more for trees has been given. This means that 25 to
30 cm of soil moisture can be removed by trees before
transpiration falls below the potential rate. Concluding
it can be said that the potential evapotranspiration
approaches the actual evapotranspiration if the soil
moisture content is above a certain level.

Because of the nature of the soil, loam and silty
loam texture and the considerable depth (> 2 m), the rather
low precipitation deficit in the summer months (table 3J,
fig 2) and the fact that the Haarts catchment is completely
forested, with more or less everywhere a rather deep
rooting zone (up till 200 cm) it is assumed that the soil
moisture content in the root zone is very rarely a factor
of limitation for the evapotranspiration. Consequently
the actual evapotranspiration should approach the potential
evapotranspiration very closely. This conclusion is
corroborated if the two hydrological years November 1973 -
October 1975 are considered. As no surface inflow or
subsurface in- or outflow took place and the storage is
the same at the beginning and end of these years, the water
balance equation can be simplified to:

Precipitation = Evapotranspiration + Discharge. From
this equation is can be determined that the potential
evapotranspiration, calculated with the formula of Haude,
deviates by only 49 mm for these two years which is only
6 percent. This difference is within the error of the
precipitation and the discharge measurements.

2.   VEGETATION AND LAND USE

The Haarts catchment is a completely forested area.
It consists of a mixed oak-beech forest, with some areas of
coniferous woodland (fir and spruce) and one small area
with Norwegian spruce and grass (fig. 3). In this figure
a distinction was made within the group of coniferous
woodland between "Old" spruce and fir, which were mostly
planted in the period 1890-1914, and the younger trees
planted shortly after World War II.

Pollen analysis of alluvial and slope deposits from
the area investigated and old maps and historical
documents supply evidence for reconstructing the
evolution of the vegetation from the Late Dryas up until
now (Firbas 1949 and 1952; Faber & Schmidt 1963; Riezebos
& Slotbook 1974; Kwaad & Mücher 1977; Riezebos &
Slotboom 1978):

1.   Treeless grassland period (Zone III. Late Dryas):
     characterized by high percentages of the non-arboreal
     pollen including *Artemesia, Helianthemum,
     Sanguisorba and Thalictrum*; the transition to the
     postglacial is shown by a decrease of the herbaceous
     species, while *Pinus* increases.

11

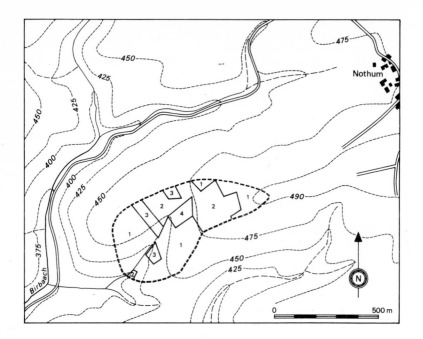

Figure 3   Vegetation map of the Haarts catchment.
1. Deciduous woodland (oak, beach)
2. Coniferous woodland
3. Young coniferous woodland
4. Grass with Norwegian spruce

2. Birch-Pine period (Zone IV; Preboreal): dominant
    species are *Betula* and *Pinus;* decrease of *Salix;*
    a threefold division is possible - Friesland
    oscillation (65 % trees) - Piotino oscillation
    (decrease in *Pinus* and increase of *Betula*;
    9235 ± 80 BP) - third phase more or less late
    Friesland oscillation.

3. Hazel-Pine-Birch period (Zone V; Boreal): rise of
    the curve of *Corylus* and of *Ulmus;* in the youngest
    parts *Pinus* has already considerably decreased.

4. Mixed Oak-Hazel-Alder period (Zone VI/VII; Atlantic):
    Quercus, Tilia and *Alnus* appeared to be increasing

5. Oak-Beech period (Zone VIII, Subboreal): at the
    beginning still important percentages of *Quercetum
    mixtum*, particularly of *Quercus* and *Corylus,* but
    afterwards a strong increase of *Fagus* and a strong
    decrease of *Ulmus.*

6. Beech period (Zone IX, Older Sub-Atlanticum): the
    dominant species is *Fagus,* pollen from which is two
    or three times more abundant than *Quercus;* the
    Alnus content is less, up until now; slow increase
    of *Carpinus;* the beginning of this zone is dated about
    100 years AD.

7. Beech-Oak-Pine-Spruce period (Zone X, Sub-Atlanticum): this zone started immediately after the *Fagus* maximum and shows a strong increase in cultivated and ruderal plants (Zone Xa; Cerealia, *Fagopyrum*, *Artemesia, Chenopodiaceae, Plantago* and *Rumex*); the sharp increase of Cerealia is dated about 1325 AD: buckwheat *(Fagopyrum)* was introduced about 1460 AD (Slicher van Bath,1962); the spectra of the subzone Xa indicate an open landscape; the boundary between subzone Xa and Xb is placed where *Picea* begins and *Pinus* increases due to the plantations after 1850 AD.

The pollen data indicate a continuous deforestation in the Oesling during the Sub-Atlantic times preceding the 13th century. In the Oesling the stage of this deforestation corresponds with the Late-Germanic period of occupation from the 8th to 12th century. In the 12th and 13th century Cerealia strongly increases due to the high cereal prices (Riezebos & Slotboom 1978). After this period the population declines (1300-1400 AD) and cereal prices fall, so that the Cerealia curve decreases and is followed by an advance in pastural farming. Also flax culture was introduced but this cultivation period ended about 1625 AD. From the 13th to the 19th century a threefold system was very common in arable farming, first year - wintercorn, second year - spring corn, and third year - fallow. Between 1500 and 1800 AD pasture was mainly restricted to the valleys, but after 1800 AD the arable land/pasture ratio decreases (Riezebos & Slotboom 1978). From about 1850 and especially after the discovery of tanning with chromic acid instead of tan from oak, Spruce was extensively planted in the Oesling. Two periods of plantation can be distinguished. The first was from 1880-1914, but this period of Spruce plantation ended when World War I began and a revival of the old tanning process took place. The second period started after World War II and in the sixties about 300 hectares per year were reforested (Faber & Schmit 1963). The importance of the evolution of the vegetation and the disturbing of the environment by man related to the regolite and slope deposits will be discussed in Part A II 4.

In general at this moment in the Oesling land use is more or less related to the topography. The flat plateau and gentle upper slopes, if not too limited in area, are used for arable crops and grass. The gently to moderately sloping valley bottoms are generally under permanent grass for grazing and hay, though recently a number of these have been planted with spruces. The steeper slopes and very small plateaus, as in the Haarts catchment, are completely forested.

3. GEOLOGY

The Oesling is a part of the Ardennes which forms, together with the "Rheinische Schiefergebirge" one of the main units of the Hercynian massifs on the European continent. The Oesling can be subdivided into three main structural units, all consisting of Lower Devonian marine sediments which were folded during the Lower Carboniferous. These three structural units are the Bastogne Anticlinorium and all have a WSW-ENE strike and a plunge, varying from a few degrees up to locally 20^0 towards the East. Sometimes locally some transversal faults occur and within each of the three main units some secondary folds are present, causing a pattern of small elongated anticlines and synclines.

The central part of the Oesling is occupied by the Eifel Synclinorium, running from the Meuse in the West towards the "Kalk Mulde" of the Eifel in the East. In the Oesling this synclinorium is called the Wiltz Synclinorium and it is composed of, from the centre outwards, Upper-, Middle- and Lower Emsian rocks (Lucius 1950a).

In the South, the central synclinorium is bounded by the Givonne Anticlinorium (fig. 4). In the western part Cambrium rocks form the core of this anticlinorium. Due to the plunge, younger rocks of the Lower Devonian (Siegenian and Lower Emsian) are found towards the East and also many secondary anticlines and synclines have been developed, resulting in a fingershaped surface pattern of older and younger rocks of the Lower Devonian.

To the North the central synclinorium is bounded by the Bastogne Anticlinorium (fig. 4), which forms the southern axis of the main anticlinorium of the Ardennes, of which the latter is called Stavelot Massif. The core of the Bastogne Anticlinorium is composed of Cambrium rocks, with outcrops in the western part (Rocroi Massif) and more to the East (Serpont Massif and Stavelot Massif). The Devonian rocks in between the first two massifs are of Gedinnian and Lower Siegenian age. East of the Serpont Massif the axis of the anticlinorium deviates from W-E towards a SW-NE direction and here in the central part of the anticlinorium the rocks are of Gedinnian till Upper Siegenian age.

In table 5a a stratigraphic division of the Devonian in the Oesling is given according to Lucius (1950a). Lucius makes a distinction between the indistinct stratified coarse "Schiefer" with rare banks of argillaceous sandstone (Upper Siegenian) and the "Schiefer" with a well developed stratification (quartzophyllades) interbedded with quartz sandstones typical for the Lower Emsian. However, new data on fossils show that it is not yet possible to make a chrono-stratigraphic distinction between the Siegenian and Lower and Middle Emsian, and a more simple division has to be made (table 5b).

The field area is situated somewhat south of the central part of the Wiltz Synclinorium (fig. 5). Here a

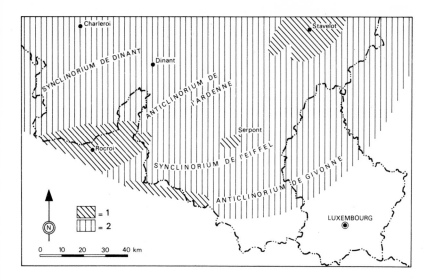

Figure 4   General view of the Paleozoic structure of a
part of Belgium and Luxembourg
1. Caledonian Massifs tectonized by Hercynian
orogeny
2. Ardennes proper
(after Fourmarier 1954)

rather simple symmetrical structure exists, resulting in
a generally undisturbed sequence of formations. According
to Lucius's (chrono) stratigraphic division (table 5a),
the catchment consists of the "Schiefer von Stolzemburg"
(E1a) with only in the northern part a very small area of
the "Quartzophylladen von Schüttburg" (E1b). Both belong
to the "Tonschiefer-Grauwacken-Quartzit" series (table 5b).

The "Schiefer von Stolzemburg" is characterized by
dark coloured phyllites and quartzophyllites, with rare
thin intercalations of sandstone. The phyllites are
massive, well compacted blue-grey "Schiefers" with sub-
parallel mica orientation, their bedding only visible at
joint faces. The "quartzophyllites" cover a regular
alternation of shales ("Tonschiefer") and sandy "Schiefer"
with a rather good cleavage. These sandy "Schiefers" are
rather fine-grained but with a gritty texture and grey
till greenish grey colours. The sandstones have a well-
cemented gritty texture.

The "Quartzophyllites of Schüttburg" are characterised
by frequent intercalations of quite massive dark brownish
quartz-sandstone, which occur interbedded with masssive
well-compacted "Schiefers" and sandy "Schiefers".

It can thus be concluded that the investigated
watershed has been built up by a sequence of rocks showing
a rather rapid alteration of compacted "Schiefers"
(shales) and sandy "Schiefers" (sandy shales) with only
very few unimportant outcrops of quartz sandstone.

Table 5 a) (left)   Stratigraphic division of the Lower Devonian in the Oesling (Konrad & Wachsmut 1973)

b) (right)   Stratigraphic division of a part of the Lower Devonian in the Oesling (Konrad & Wachsmut 1973)

a)

Unter-Devon			
Emsien (E)		Oberes (E3)	Schiefer von Wiltz (E3) Quartzit von Berlé (q)
		Mittleres (E2)	Bunte Schiefer von Clerf (E2)
		Unteres (E1)	Obere Abteilung (E1b): Quarzsandstein und Quarzophylladen, bezeichnet als "Quarzophylladen von Schüttburg" Untere Abteilung (E1a): Schiefer mit guter Schichtung, Quarzophylladen und seltene Bänke von Quarzsandstein, als "Schiefer von Stolzemburg" bezeichnet.
Siegenien (Sg)		Oberes (Sg3)	Oberes Siegenien im allgemeinen (Sg3): Undeutlich geschichteter Grobschiefer (Sg3) mit seltenen Bänken von tonigem Sandstein. An der Basis lokale Fazies von Dachschiefer (Sg3a) Im äussersten Norden des Oeslings mehr sändige Fazies (Sg3s)
		Mittleres (Sg2)	Sandstein, oft fossilreich, und sandiger, kompakter Schiefer
		Unteres (Sg1)	Pylladen und Quarzophylladen

b)

Oberes Emsien	Schiefer von Wiltz (= E3) Quartzit von Berlé)= q
Unteres Emsien	Bunte Schiefer von Clerf (= E2)
Oberes Siegenien	Tonschiefer-Grauwacken-Quartzit-Wechselfolge (= E1b + E1a + Sg3 + Sg3a + Sg3s)

16

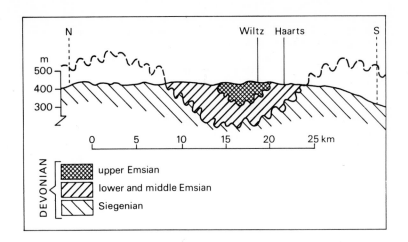

Figure 5   Schematic cross-section of the main structural
            units in the Southern Oesling (after Bintz
            et al 1973).

The shales and sandy shales have a slaty cleavage and
are very compact impermeable rocks, disintegrating
exclusively by weathering along the cleavage planes. In
most cases these cleavage planes are present in a latent
stage within a depth of one meter from the surface of the
rock (Lucius 1950a).

Joints occur frequently and two preference directions
can be distinguished (60° and 150°). These are longitudinal
and transversal joints of which the latter predominates
(80%) (Bintz 1964). In spite of the frequency of the
joints, their influence is very restricted at some depth.
The joints in the (sandy) shales are mostly closed within
one meter, but sometimes, predominantly in the quartz
sandstones, open joints exist up to a depth of 10 meter.
This, together with the very small area of the sandstones,
explains that the Lower Devonian rocks in the Oesling form
an extremely small water reservoir (Lucius & Bintz 1960;
Bintz 1964) in which water circulation is only possible
within the few upper meters of these rocks (see also part
A II 5).

4.   GEOMORPHOLOGY

4.1   Landscape history

The Oesling forms part of an old planation surface,
which was uplifted and then dissected by rivers during the
Upper Pliocene and Pleistocene. At this moment several
opinions exist about the formation of the present relief
of the Oesling.

Lucius (1940, 1950a, b) postulated the following
evolution of the Oesling. At the end of the Lower
Carboniferous, during the Hercynean orogeny, a planation

surface developed ("post-Hercynean peneplain") in the
Ardennes. During Triassic times a transgression took place
and from then until the Lower Cretaceous Mesozoic
sediments were deposited on the planation surface. In
Lower Cretaceous times erosion took place and the Mesozoic
rocks of the Oesling were partly removed. In the central
part of the Ardennes, in Belgium, the old pre-Triassic
planation was stripped of sediments and another planation
surface developed (pre-Maestrichtian surface) (Lefevre
1938; Macar 1954). During the Senonian, the Oesling was
again completely covered by the sea, but from the Eocene
onwards subaeric erosion and denudation could take place.
A strong uplift at the end of the Pliocene was very
important for the development of the relief of the Oesling.
The southern part of Luxembourg, the Gutland, was uplifted
to a much lesser extent than the northern part. This
differential uplift gave rise to a flexure at the border
of the Oesling and Gutland. The Mesozoic cover of the
Oesling was removed during the Pleistocene, resulting in
the exposure of the original pre-Triassic peneplain. This
surface has not been lowered appreciably since that time
(Lucius 1950a). During the Pleistocene, the old planation
surface was uplifted and consequently dissected by fluvia-
tile erosion, but the character of the planation surface
has nevertheless been preserved. However, in the area of
the Wiltz Synclinorium, to which the investigated area
belongs, and which consists of relatively soft shales, the
planation surface was more fully dissected, resulting in a
relief of many small plateaus, from which long, gentle
slopes lead to steeply incised valleys.

In conclusion, according to Lucius' view the Oesling
is a dissected pre-Triassic planation surface which has
hardly been remodelled by denudational processes during
the Pleistocene. Many authors, however, have recognised
in the Ardennes several planation surfaces dating from
Tertiary periods (Baeckeroot 1942; Gullentops 1954;
Macar 1954; Alexandre 1957; Stevens 1959; Pissart 1961).
As far as the Oesling is concerned, the existence of a
pre-Triassic peneplain has been rejected by Piket (1960)
and Verhoef (1966). The former recognised an Oesling
peneplain at an altitude of 500-520 m, characterised by
thick paleosols dating from the Lower and Middle Pleistocene. In addition two pediment-like planation surfaces
from the Upper Pliocene were recognised.

Another view of the development of the relief of the
Oesling was proposed by Hermans (1955). According to this
author, following the theory of denudational altiplanation
of Bakker (1948) and De Bethune (1948), periglacial
denudation and erosion have considerably influenced the
Oesling relief. The Oesling relief is seen as a rolling
landscape, which was lowered during the Pleistocene, so
that the original pre-Triassic planation surface does no
longer remain and the rather undissected parts of the
Luxembourg Ardennes are covered by solifluctional deposits.
In areas with a high drainage density and steep slopes,
like the southern part of the Oesling, the solifluction
layer is less ubiquitous and only occurs locally.

Finally it can be concluded that the area of the Wiltz Synclinorium has been remodelled during the Pleistocene, creating a relief of small plateaus and steeply incised valleys.

4.2 Late Quaternary landscape evolution in the Birbaach catchment

A number of recent investigations in the Birbaach catchment and adjacent areas (Riezebos & Slotboom 1974; Kwaad & Mücher 1977; Imeson 1977; Kwaad 1977) have thrown light on (sub) recent evolution of the superficial deposits. In this part of the Oesling many small plateaus exist, from which long, gentle upper slopes lead to steep lower slopes with maximum valley-side slope angles over 30°. With the exclusion of the small plateau summits, in all of the geomorphic units slope deposits can be recognised in many places (fig. 11). In general it appears that at least two slope deposits can be recognised. An older one, consisting of Würm solifluction material, and a more recent one, formed when the area was brought under cultivation.

A relatively old slope deposit, amongst other things, is characterised by an argillic horizon, a lack of pollen, and volcanic minerals from the Laacher See eruption during the Allerød like brown-green amphibole, sphene and clino-pyroxene, with the exception of the few upper centimeters, and the almost absence of charcoal (= II horizons of fig. 11; see also Imeson & Jungerius 1974). The age of formation of the argillic horizon has to be older than Sub-Atlantic, because it is buried by a secong, younger, slope deposit of Sub-Atlantic age. From the fact that almost none of the illuviation cutans have been disturbed by gelifraction or swelling or shrinking, it can be concluded that an interglacial age is unlikely and that the age of the argillic horizon might be Atlantic and/or Subboreal and that it has developed in a Würm mass-wasting deposit (Kwaad & Mücher 1977, 1978).

This older type of slope deposit is covered by a younger slope deposit, which could be distinguished in forested areas from the underlying material amongst other things, by a lack of illuviation cutans, the presence of pollen and of volcanic minerals from the Laacher See eruption and an abundance of charcoal. This younger slope deposit shows the same pollen stratification as the peat and alluvial deposits in the area (Riezebos & Slotboom 1974). Although the identification of slope deposits by pollen analysis is still in a rather initial stage, it was con-cluded that this stratification reflects the age, the vegetation and the climatic conditions during deposition. It could be established by palynological investigations by Slotboom (see Kwaad & Mücher 1977, 1978, and appendix II) that in this area the upper part of the profiles, above the argillic horizons, accumulated during the latter part of the Sub-Atlantic (zone Xa and Xb of Firbas (1952)). The presence of buckwheat (Fagopyrum) in this deposit, which has been introduced into the Oesling about 1460 AD, indicates that this younger slope deposit was formed after this time. The ratio aboreal-nonarboreal pollen and the spectra of zone Xa indicate an open landscape (fig. 31, table 59). The upper part of this younger slope deposit with a high content of arboreal pollen (30 → 80%) represents zone Xb and reflects the expansion of the forested area after 1800 AD, due to reafforestation (*Picea* plantations) (Kwaad & Mücher 1977).

On the steep lower slopes beneath the slope deposit with the argillic horizon, another, probably older mass-wasting deposit having a grèze-litée like character occurs (III horizons of fig. 11) (Guillien 1951, 1964). This layer, which consists of gravel and stones (81 %) is almost purely mechanically disintegrated bedrock as a result of physical weathering in a glacial period.

It can thus be concluded that during the Late Pleistocene at least one slope deposit has accumulated. During the Early Holocene until the Subatlanticum a period of landscape stability prevailed, resulting in the formation of soils with an argillic horizon all over the area except probably on the smaller plateau summits. Due to deforestation and agriculture, during the Middle Ages soil erosion occurred and soil profiles were (partly) truncated in many places. These changes in the environmental conditions, which occurred until about 1800 AD resulted in the formation of a second slope deposit. From then until now another period of landscape stability dominated the forested areas, which is mirrored in the low sediment load of the rivers and the formation of the Umbric Dystrochrepts in the Haarts and adjacent areas.

4.3   Processes responsible for recent landscape evolution in the Birbaach catchment

4.3.1   General

From the discussion above it is clear that the Birbaach area has been subjected to periods of relatively intense erosion, but that today erosion processes are under forest relatively unimportant and that a period of relative stability exists.

It is logical to assume that the erosion associated with the Medieval forest clearance and buckwheat introduction was similar to that found today either on arable land in the neighbouring areas or with processes of accelerated soil erosion described today in the soil conservation literature (Ellison 1944a, b; Horton 1945; Hudson 1971). Since such processes are well documented and are now irrelevant because of the fact that the Haarts catchment is completely wooded, they will not be discussed further. Only those processes operating today under the forest in the watershed will be given.

4.3.2   Processes in the Haarts catchment under forest

From general observations and earlier work it is evident that the processes operating today in the Haarts catchment reflect (1) the nature of the forest, and (2)

the drainage conditions of the soil. Before considering
their distribution in the Haarts catchment, reference will
be made to process studies undertaken in the area.

From hydrological and geomorphological investigations
(Imeson & Jungerius 1974; Imeson 1977; Kwaad 1977;
Verstraten 1977) it could be established that in the
Haarts catchment no soil erosion by Horton overland flow
appears. Porosity, permeability and aggregate stability
are all high for the soils in the drainage basin. The
only recorded overland flow in this area is an insignificant
but frequently operating movement over short distances of
water through and over the litter layer. The volume of this
litter flow is always small in relation to the amount of
rain measured beneath the forest canopy, even during
heavy storms (Imeson & Jungerius 1974).

The most important process for the material transport
and sediment supply into the river is splash erosion
(Imeson 1977: Kwaad 1977). However, under forest this
process is impeded by the protection of the ground by
litter and low vegetation. In the Haarts catchment
significant splash erosion only occurs when mineral soil
is exposed. It is significant that most of the exposed
soil is found within 3 m or so from the river and that at
least 95 per cent of the bare soil is in one way or another
the result of animal activity, predominantly burrowing by
moles and voles (Imeson 1976). For several individual
storms it has been established that the sediment supply
into the river is completely controlled by splash erosion
on bare land along the channel.

Also the influence of other burrowing animals, like
earthworms, especially in the moist areas near the river
and the effect of needle-ice growth on material transport
has to be considered. Unfortunately no data are
available yet. The effect of larger animals in the watershed in displacing material appears to be quite small
($34 \text{ m}^3/\text{km}^2$), although in the case of wild pigs this is
sometimes spectacular (Imeson 1977).

The output of suspended solids measured in the stream
channel is low. Suspended solids concentrations are
generally less than 10 mg/1, unless periods of rain coincide
with high discharges or critical discharge is exceeded and
coarse bed material is transported. This latter usually
occurs once or twice a year. Suspended matter concentrations
may also be higher (30-40 mg/1) following frost. Also
relatively high suspended matter concentrations are
measured during small periods of rain. Although it is not
yet possible to give an accurate estimation of the sediment
output all over the year in the drainage basins, it seems
to be low (Imeson 1977). This assumption is also
corroborated by measurements of sediment output in the
adjacent Upper Birbaach catchment (Imeson & Jungerius
1974). So it can be concluded that today in the Haarts
catchment landscape stability prevails with dominating
soil formation and only slow colluviation, resulting in
the formation of Typic or Umbric Dystrochrepts (see A II 6).

As far as the Haarts watershed is concerned, it is important to emphasize that considerable variation exists in the pattern of erosion. Particularly significant is the fact that: (1) Processes are concentrated along a 5 m belt of land along the river channel where seepage promotes a rich microfauna, considerable animal activity, the development of needle-ice, conditions of positive porewater pressure, and also a relatively low aggregate stability. The steep slopes near the area are also poorly covered with litter and are exposed to splash erosion throughout the year. (2) Under coniferous woodland the soil is completely covered by needles and neither opportunity for splash erosion, nor any transport of sediment attached to the leaves occurs.

5. RUNOFF CHARACTERISTICS OF THE HAARTS CATCHMENT

5.1 Introduction

The runoff from the Haarts catchment is considered in some detail. This is because the reaction times of the various runoff components enable some conclusions to be drawn concerning the source of the runoff and its residence time in the catchment. This is in addition to the obvious usefulness of runoff records in calculating the rate of chemical denudation. The data for this chapter are derived from a paper by Imeson and Verstraten (1979).

The Haarts catchment is drained by a single stream, without tributaries, 250 m long. In addition to the springs shown in fig. 1, which supply most of the water, the river receives much of its discharge from seepage zones along the steeply incised channel (fig. 11). The profile of the river is steep (average slope 7.5°) and most of the channel is cut into bedrock. There are only a few gravel accumulations of relatively small extent so that there is little storage opportunity in the channel. In spite of the steep slope, tracer (pyranine) experiments (Imeson & Verstraten 1979) showed that over the irregular channel bed water requires, under most flow conditions, at least an hour to pass from the highest source to the catchment outlet. During winter months and early spring the river extends about 100 m up valley to the spring indicated at point A (fig. 1). Here the channel is not incised and in places the water flows underground beneath relatively recent colluvial deposits (under dry valley).

The catchment gauging station (fig. 1) installed in May 1973, consists of a 90° V-notch weir and an Ott water level recorder. Between 1974 and 1976 the discharge was also measured 200 m upstream (fig. 1) with a 45° V-notch weir and Munro water level recorder.

5.2 The general pattern of runoff

The discharge from the Haarts catchment is remarkably seasonal, with only 5-20 % of the annual discharge occurring during the summer half of the year. The seasonality is partly a reflection of the yearly cycle in (potential) evapotranspiration (fig. 2) and of the lack of an important groundwater reservoir.

Summer periods of rainfall produce relatively small hydrograph peaks which quickly recede. Exceptionally large and intense periods of rainfall are required, such as the 3 cm that fell on August 11th, 1975, to produce discharges approaching those reached during the winter. The quick recession of the summer peaks indicates that relatively little increase in catchment storage is associated with them. Because they are well defined, they can very easily be separated into components of quick and delayed flow (see Hewlett & Hibbert 1967). Quickflow is generally small in relation to the amount of catchment precipitation, being seldom equivalent to more than the rain received by an area four times as large as the channel surface. Due to the variable nature of summer rainfall it can be either insignificant or very important in any single month. Generally only about 5 to 15 per cent of the monthly discharge between May and October inclusive is quickflow, but in some months it has reached 85 per cent. Of course most quickflow occurs for only a few days each month. In the summers of 1974 and 1975 respectively 26 and 45 per cent of the runoff was quickflow.

During the winter half of the year, the runoff behaves quite differently. Periods of rain or snowmelt produce, about 12 hours later, rises in the hydrograph which are far more gentle than the summer ones, and which often persist for several days at a relatively high level, before receding relatively slowly over a period of weeks to a level of delayed flow much higher than that observed in summer. Any direct response of the discharge to rainfall is so small in relation to the total discharge that it can not be determined accurately from the catchment runoff record. This was not the case during certain periods in the spring and autumn when the winter delayed flow component was relatively small. Periods of heavy rainfall then produce double hydrograph peaks, comparable to those described by Weyman (1974), the first essentially similar to those observed in the summer, and the second resembling the winter ones. The distinction between quick and delayed flow in the winter is sometimes not obvious. The separations were made according to an arbitrary but consistent method of separation (Hewlett & Hibbert 1967).

The proportion of the runoff which is quickflow is much more constant in winter than in summer. For the months characterised by the winter runoff response, between 50 and 75 per cent of the monthly runoff was quickflow in 1973/74 and 1974/75. It is remarkable that for both of these winters a similar percentage of runoff was quickflow (62 per cent), even though the winter of 1974/75 was much wetter and the river discharged three times as much runoff.

On the basis of the general runoff characteristics a division of the year into summer and winter periods is likely to be useful in an attempt to interpret the various

23

runoff components in terms of their relevance for the
water chemistry.

5.3   Rainfall-runoff relationships

   Parameters of a precipitation-runoff model can be
derived from discharge hydrographs of a catchment. The
fundamental idea is that the discharge hydrograph of an
area shows the hydrologically characteristic properties
of that area and will thus yield the parameters of the
model (De Zeeuw 1973). On its way to the catchment outlet,
water (effective precipitation) passes through various
forms of storage. Several "reservoirs" can be postulated
where water is temporarily stored, namely the surface
reservoirs, soil moisture, groundwater and channel
reservoirs (for further details see Krayenhoff van de
Leur 1973; De Zeeuw 1973). This "reservoir effect" was
indicated by Lyshede (1955) and the hydrograph can be
described as a sum of exponential functions that are
expressed as a number of reservoirs (linear or non-linear).
Several investigations have suggested that either the
entire recession curve, or segments of it can be
expressed in individual exponential functions. Hall (1968)
showed that the major point of contention, being either
an aquifer or another source of base flow, has a linear
or non-linear response. The general non-linear equation
is $Q_t = Q_o (1+\alpha t)^{N/(1-N)}$, in which Q_t is discharge at time
t, Q_o is initial discharge at time t = o, α is the
recession constant and N is a constant (N $\neq$ 1). Drogue
(1972) reviewed the effect of various linear and non-
linear reservoirs and concluded that spring discharges
are satisfactorily represented by a non-linear curve of
the form

$$Q_t = \frac{Q_o}{1 + \alpha t^N}$$

Many investigators (Forkasiewicz & Paloc 1965; Boussinesq
1905) showed that the non-linear case could just as
easily and accurately be represented by the linear form,
using several linear reservoirs superimposed. In a linear
reservoir the outflow rate is proportional to dischargeable
storage. The flow and continuity equations for a linear
reservoir are (De Zeeuw 1973):

   flow equation: $Q = \alpha S$

   continuity equation: $P_e = Q + \dfrac{dS}{dt}$

where Q = discharge per unit surface area in mm/day
      S = storage per unit surface area in mm
      α = reaction factor in day^{-1}
      P_e = effective precipitation per unit surface area
         in mm/day

A combination of equations (1) and (2) results in the equation:

$$Q_n = Q_{n-1} e^{-\alpha(t_n - t_{n-1})} + P_{e,n} (1 - e)^{-\alpha(t_n - t_{n-1})} \qquad (3)$$

where Q_n is the discharge and $P_{e,n}$ the depth of P_e during interval t_{n-1} to t_n. For periods without effective precipitation in which no recharge takes place equation (3) is reduced to

$$Q_t = Q_o e^{-\alpha t} = Q_o e^{-\alpha(t - t_o)} \qquad (4)$$

The value of α can be found by plotting the hydrograph on semi-log paper. For detailed accounts of the method see Van te Chow 1964 and De Zeeuw 1973.

The reaction factors α of the various reservoirs, ie. the absolute value of the slope of the recession line plotted on semi-log paper canalso be computed as follows (De Zeeuw 1973):

$$Q_2 = Q_1 e^{-\alpha(t_2 - t_1)}$$

$$\log Q_2 = \log Q_1 - \frac{\alpha(t_2 - t_1)}{2.30}$$

$$\therefore \quad \alpha = 2.30 \frac{\log Q_1 - \log Q_2}{t_2 - t_1}$$

The total discharge can be obtained by summing the outflows from the various reservoirs.

The discharge at any instant time, t_n from t_o, is:

$$Q_{tn} = Q_o' e^{-\alpha' t} + Q_o'' e^{-\alpha'' t} + Q_o''' e^{-\alpha''' t} \quad \text{etc.}$$

Attempts to reconstruct hydrographs using a reservoir model failed because in practice the effective precipitation proved impossible to estimate accurately and secondly because the contributing areas of the various reservoirs were probably temporarily and spatially variable. However, the recession constants of the hydrograph components give some insight into the drainage of the catchment.

5.4 Runoff on the catchment slopes

At a number of sites, water moving laterally downslope was intercepted, at the surface and at various depths down to 3 m. Infiltrating water was also collected. The method of measurement involved the excavation of large pits and the installation of plastic plates and perforated pipes (fig. 6), basically in the same way as described by Whipkey (1964).

Figure 6   PVC drains in profile Haarts 1

From the measurements it can be concluded that during the observation period Horton or saturated overland flow did not occur, except through the litter layer (see also Imeson & Jungerius 1974). Further it is likely that in the extremely permeable soils water percolates predominantly downwards with little lateral movement. To a depth of 2.5 - 3 m, throughflow was insignificantly small. Infiltration was very rapid and ceased soon after rainfall. During the summer, no water percolated through the soil profile to the weathering zone.

From field observations it would seem that lateral transport of water on the steep catchment slopes took place in the weathering zone through the regolith or possibly a grèze litée-like slope deposit characterised by an extremely high porosity and a gravelly texture. This gravelly layer is of widespread occurrence, but it is not found everywhere indicating that the drainage of the slopes is not completely uniform. That runoff from the catchment is supplied from the weathering zone, and from very shallow joints in the shales, is also indicated by the water temperature at springs and seepages. The annual range in water temperature in relation to the surface temperature suggests that the water has been in contact with rock and weathering products at a depth of 3 - 4 m (Heath & Trainer 1968).

5.5 The major components of runoff

5.5.1 General

From the general consideration of runoff on the slopes and in the catchment stream the following major runoff components can be recognised and provisionally physically interpreted:

	Interpretation
1. Summer delayed flow	Water probably supplied from the entire catchment area and entering the stream at springs and seepages along most of the channel.
2. Summer quickflow	Channel precipitation, including rain falling on partial areas very close to the river channel
3. Winter delayed flow	A continuation of the drainage associated with (4) but water supplied from more gently sloping and more remote areas, as well as from some slope deposits along the channel
4. Winter quickflow	Partly channel precipitation, but for the most part runoff from permeable soils on steep slopes

The double hydrograph peaks of the spring and autumn are not treated as a special case. Only the summer quick-flow component bears no relationship to the chemical parameters measured at the various spring sites described later.

5.5.2 Summer delayed flow

An abundance of literature exists on the analysis of base flow recessions (see for example Barnes 1919; Hall 1968; Waugh 1970; Singh & Stall 1971, and Drogue 1972). Since no recharge of water was observed on the catchment slopes, the delayed flow (base flow) might be expected to decrease gradually throughout the summer, as the amount of storage decreases according to an exponential or other function (Drogue 1972). An examination of this relationship at the lower gauging station was hindered by the fact that some recharge and storage of water seemed to be occurring close to the river, during wet summer periods since the level of base flow increased for a few weeks following heavy rainfall. This effect is most pronounced when the base flow is very low, as in 1976. When individual base flow recessions are considered for storms followed by long dry periods, these could be described by simple exponential curves of the form:

$Q_t = Q_o e^{-\alpha t}$, where Q_t is the discharge at time t, Q_o is

the initial discharge at time t = o and α is the
recession constant (reaction factor) in days^{-1}. The
value of ($e^{-\alpha}$) was in all cases either 0.97 or 0.98 and
$\tau \equiv (\frac{1}{\alpha})$ varied between 32 and 44 days.

In contrast, at the upper gauging station, a good
relationship could be found for the summer as a whole,
the discharge at this site did not seem to reflect
changes in storage produced by summer floods. The value
of τ for 1975 was 86 days and for 1976 64 days. Although
the curve $Q_{t_2} = 0.034e^{-0.156t}$ provided a very good fit to
the date ($r^2 = 0.98$), the relationship is not linear.
Thus for September τ reaches a value of 209 days. A non-
linear response particularly at the end of the recession
is not unusual (Hall 1968; Nutbrown & Downing 1976).

5.5.3 Summer quickflow

The variable occurrence of quickflow during the
summer has already been mentioned. In table 6 monthly
quickflow amounts and percentages are shown together with
the amount of rainfall recorded at Berlé.

Table 6 Discharge of summer quickflow at the Haarts
gauging station during 1974 and 1975

month	delayed flow (mm)	quickflow (mm)	quickflow (% of total flow)	rainfall at Berlé (mm)
April '74	8.4	1.3	13	21.9
May	5.6	2.4	35	40.5
June	2.3	1.7	42	74.1
July	2.3	0.2	9.2	63.6
August	1.6	0.1	6.4	42.8
September	3.9	2.7	40	128.3
May '75	18.4	2.5	12	38.1
June	5.2	0.3	5.7	65.7
July	2.1	0.3	13	51.4
August	1.9	11.2	85	51.0
September	1.9	0.4	18	89.2
October	1.9	0.1	4.0	16.8

Virtually all periods of rain greater than about 5 mm
produce the characteristically sharp peaked hydrographs
which recede at first with a reaction time (τ) of between
3.5 and 4 hours. The effect of smaller amounts of rain is
difficult to discern from the hydrograph. The peaks are
formed by only very small percentages of the total rainfall

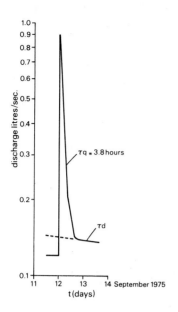

Figure 7

Peak flow during summer periods

received by the catchment, as the peak in fig. 7 indicates. This was produced by 11 mm of rain. If the effective precipitation in estimated as about 7 mm, then the volume of runoff is equivalent to only 1.2 per cent of this (0.08 mm). On the other hand the amount of runoff is equivalent to all of the rain falling on a zone 6.8 m wide along the river. In most cases this equivalent zone is only 3 or 4 m wide. This partial area approach would seem to offer the most acceptable explanation for the observed hydrographs.

5.5.4 Discharge during the winter period

The hydrographs occurring during the winter months, as mentioned earlier, begin about 12 hours after rainfall or snowmelt and produce peak flows which persist at a fairly constant discharge for generally between 2 and 5 days. These peaks recede with time constants of between 4 and 7 days. When plotted on semi-logarithmic paper, the hydrographs reveal a number of straight line segments but only in about 50 per cent of the cases is there a distinct change in gradient which might be associated with the transition from quick to delayed flow (fig. 8). In other cases the break of slope associated with this transition is indisctinct so that such a separation is rather arbitrary. Nevertheless delayed flow and quickflow have been separated to determine the various time constants using the procedure described by Singh & Stall (1971). The application of this procedure is indicated in fig. 9. The recession of the quickflow component usually has a time

29

Table 7 Time constants for winter hydrograph rises*

period	begin date	τ (days)			Q (l/sec)	
		τ_d	τ_q	τ_p	Q_{Min}	Q_{Max}
1	5.2.'74	15.8	2.03	6.12	0.60	6.8
2	10.2.'74	27.18	2.68	8.23	0.68	6.8
3	14.3.'74	21.23	3.4	6.6	0.73	6.0
4	16.12.'74	7.19	1.35	2.98	8.0	26
5	26.12.'74	10.96	1.04	4.43	4.2	8.8
6	15.1.'75	49.4	0.89	4.98	2.6	6.2
7	27.1.'75	6.52	1.34	4.14	8.0	15.8

* for explanation see text

constant of between 0.9 and 1.3 days but values as high
as 3 days sometimes occur, generally when rainfall has been
recorded during the period of peak discharge. Seven
hydrograph peaks were examined in detail and the time
constants calculated for the quickflow component (τ_q), the
delayed flow (τ_d) and the peak as an entity without
separation (τ_p). These values are shown in table 7,
together with values of the levels of delayed flow at the
beginning of the rises (Q_{min}) and the maximum peak discharges attained. Six of the hydrographs are shown in
fig. 8.

No clear relationship could be found between the
peak discharge and the time constants of the recessions,
although the higher peak discharges do tend to associate
with lower values. The more rapid recessions were also
associated with higher levels of delayed flow. When the
time constants of the delayed flow are considered, the
shortest times occur after hydrographs having the highest
peak and delayed flows and the shortest quickflow time
constants. It is interesting that the shortest reaction
times occur during the winter months of December and
January and that these lengthen as the winter period draws
to an end.

The relatively gentle slope of the rising limb of the
winter hydrographs is interesting. In many cases this is
not steeper than the recession and it also plots as a
straight line on semi-logarithmic paper (τ average 1.35
days). The exact value of τ reflects the precipitation
intensity and infiltration rate. In general it is about
a factor 2 greater than the decay during December,
January and February.

The main conclusion to be drawn from the analyses is
that the rates of recession to some extent reflect the
level of discharge in the river, so that a range of time
constants is more appropriate than a single figure. For
periods longer than 10 days delayed flow recessions are

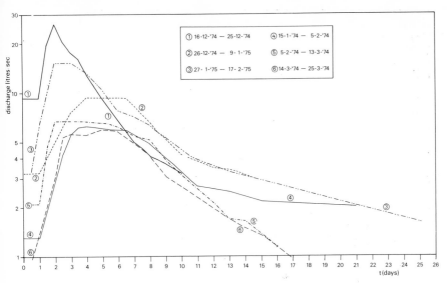

Figure 8　Six hydrographs during winter period

non-linear in contrast to the quickflow recessions which
show a reasonable amount of linearity.　It must be
remembered though that both the quickflow and delayed flow
recessions reflect the release of water from soils and
slope deposits.

6　SOILS, THEIR DISTRIBUTION, GENERAL CHARACTERISTICS AND CLASSIFICATION

6.1　General

In the Haarts catchment soils have developed on low-
grade metamorphic shales and in slope deposits.　The
general soil pattern is closely related to the four main
geomorphic units.　This pattern is also a function of the
intensity of past erosion and of the amount of accumulation
of soil material within these units.

In the Haarts catchment the following soils profiles
can be distinguished (fig. 10 and 11, table 8, see also
Kwaad & Mücher 1977):

1)　On the very small plateau summits (slope angle < 2°),
soils with an (O) -A1-R or an (O)-A11-A12-R profile
on shales;

2)　On the gentle upper slopes (slope angle $2-11^{o}$), soils
with (O)-A-B-IIBtb-IIC profiles in slope deposits;

3)　On the steep lower slopes (slope angle $15-25^{o}$), soils
in the slope deposits with an (O)-A-B-IIC or an (O)
-A-B-IIBtb-IIC profile.　The soils with an (O)-A-B-IIC
profile are only found in the higher parts of this unit;

31

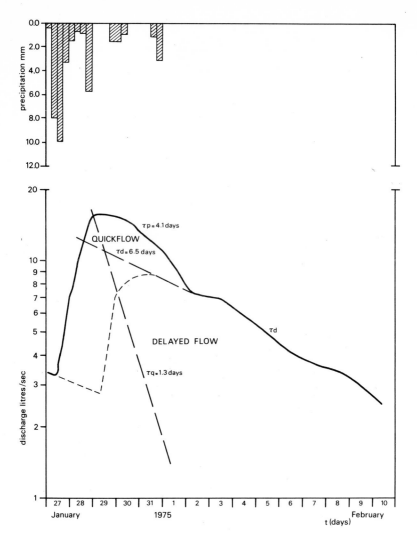

Figure 9 Example of hydrograph separation

4)   In the dry upper valley, soils with an (O)-A-C-IIBtb-
      IIC profile in slope deposits.

      Although various slope deposits have been identified
by field, palynological and micromorphological investigations (see also A.II.4), all of these deposits on the
valley slopes have a number of characteristics in common.
They are all derived from the shales with only a very
small component of quartzitic sandstone and are
characterised by a very high gravel and stone content, a
loamy texture and a similar mineralogical composition as
far as the silt, sand, gravel and stone fractions are concerned. Only the slope deposit forming the upper part of
the soil profile in the dry valley shows a somewhat
different habitus by way of its colour (higher organic

32

Table 8   View of the soils in the Haarts catchment and their classification

GEOMORPHIC UNIT	SOIL PROFILE	SOIL CLASSIFICATION	
		SOIL TAXONOMY (1975)	FAO/UNESCO Soil Map of the World:Legend (1974)
	A1 10-17 cm thick; 7.5-10 YR 3/2; B.S. < 50 % R1 15-20 cm " ; R2	LITHIC UDORTHENT	DYSTRIC REGOSOL
	A11 10-17 cm thick; 7.5-10 YR 3/2; B.S. < 50 % A12 2-5 cm " ; 7.5-10 YR 4/4; B.S. < 35 % R1 15-20 cm " ; R2	LITHIC UDORTHENT	DYSTRIC REGOSOL
SMALL PLATEAU TOPS (0-2°)	A1 18-22 cm thick; 7.5-10 YR 3/2; B.S. < 50 % R1 15-20 cm " ; R2	LITHIC UMBRIC DYSTROCHREPT	RANKER
	A11 18-22 cm thick; 7.5-10 YR 3/2; B.S. < 50 % A12 2-5 cm ; 7.5-10 YR 4/4; B.S. < 35 % R1 15-20 cm " ; R2	LITHIC UMBRIC DYSTROCHREPT	RANKER
GENTLE UPPER SLOPE (5-8°)	A1 10-12 cm thick; 7.5-10 YR 2.5/2; A3 15-18 cm " ; 7.5 YR 3.5/3; B 10-12 cm " ; 10 YR 5/5; } B.S. < 35 % IIBtb 20-22 cm thick;2.5Y 7/3+7.5YR 3/6; IIC+R1 2-10 cm " ;2.5Y 7/3+7.5YR 5/6; R1 15-25 cm thick; R2	THAPTO-HAPLUDULTIC UMBRIC DYSTROCHREPT	DYSTRIC CAMBISOL OVERLYING A BURIED ACRISOL

STEEP LOWER SLOPE (upper convex part; 15-18°)	A1 5-9 cm thick; 7.5-10 YR 3/2; A3 18-22 cm " ; 7.5 YR 4/3-4/4; ⎱ B.S. < 35 % B 14-18 cm " ; 10 YR 5/6; ⎰ IIC+R1 20-30cm" ; 2.5 Y 10 YR 5/6;	UMBRIC DYSTROCHREPT	DYSTRIC CAMBISOL
STEEP LOWER SLOPE (lower part; 19-25°)	A1 5-9 cm thick; 7.5-10 YR 3/3; (A3)20-25 cm " ; 7.5 YR 4/3-4/4; ⎱ B.S. < 35 % B2 25-45 cm " ; 7.5 YR 5/5; ⎰ B3(+R1)40-55 cm thick; 7.5 YR 5/4; IIB2tb 85-95 cm " ;10 YR 5/6; B.S. > 60 % IIC+R1 20-30 cm " ; 10 YR 5/8; B.S. > 60 % III?R1+C 20-30 cm " ; IV?R1 IVR2	THAPTO-HAPLUDALFIC UMBRIC DYSTROCHREPT	DYSTRIC CAMBISOL OVERLYING A BURIED LUVISOL
DRY VALLEY	A1 2-3 cm thick; 7.5 YR 3/2; B.S. < 50 % C 45-55 cm " ; 7.5 YR 3/4; ⎱ B.S. < 35 % IIBtb 40 cm " ;10 YR 5.5/6; IIBtb.R1 > 20 cm thick; 5YR5/8+ 10 YR 6/3; B.S. > 60 % A1+C > 50cm A1+C > 50cm	MOLLIC HAPLUDALF THAPTO-HAPLUDALFIC UDORTHENT	ORTHIC ACRISOL DYSTRIC FLUVISOL OVERLYING A BURIED ACRISOL

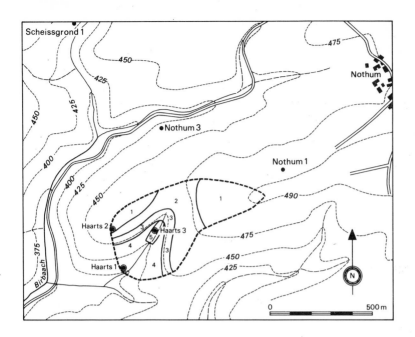

Figure 10   Soil map of Haarts catchment
1. Complex of Lithic Udurthents and Lithic
   Umbric Dystrochrepts (on shales)
2. Thapto-hapludultic Umbric Dystrochrepts (in
   slope deposits)
3. Complex of Umbric Dystrochrepts and Thapto-
   hapludalfic Umbric Dystrochrepts (in slope
   deposits)
4. Thapto-hapludalfic Umbric Dystrochrepts
   (in slope deposits)
5. Complex of Mollic Hapludalfs and Thapto-
   hapludalfic Udorthents (in slope deposits)

carbon content) and its much lower content of gravel and
stones (see appendix II, profile Haarts 3).

     The buried argillic horizon, developed in an older
(Würm) slope deposit, has been identified by means of
micromorphological investigations. Although this horizon
could be recognised in the field in deep pits, it was
impossible due to the high content of gravel and stones in
the two slope deposits, to recognise or to reach this
horizon by augering in field. Therefore only the distribution of the soils described under 1),of the cambic
horizon in 2) and 3), and of the colluvial deposit in 4)
could be established in the field with a sufficient number
of observations. These observations allowed the following
conclusion to be made: a) on the small plateau tops only
shallow soils with an A1-R or an A11-A12-R profile occur;
b) on all parts of the valley slopes soils with a cambic
horizon have developed and c) in the small upper dry
valley a colluvial deposit with a very thin A1 horizon is
found. To establish the presence or absence of the

35

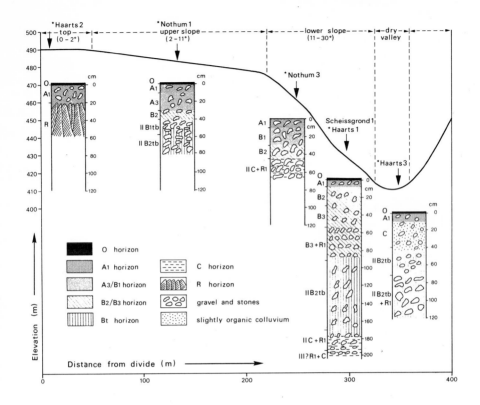

Figure 11　Schematic representation of soil profiles on
　　　　　the various landscape units (sketches are given
　　　　　of profiles with asterisk)

argillic subsurface horizon, several pits were dug. It
could be established with a relatively small number of
observations, that in the older slope deposit, which
exists all over the area except on the small plateau tops,
an argillic horizon has been developed. The only exceptions
are on some higher convex parts of the steep lower slopes,
where perhaps truncation of the argillic horizon by
erosion has been more effective and no remnants of this
horizon can be detected (A-B-IIC profile), although the
formation of a cambic horizon in the older slope deposit
cannot be excluded.

　　　Finally it has to be mentioned that in adjacent areas
in similar positions in dry valleys sometimes soils with an
O-C-IIBb-IIIBtb profile have been recognized (Kwaad &
Mücher 1977).

6.2　Soils on the plateau tops

　　　The soils on the small plateay summits are shallow
and have been developed on the low-grade metamorphic shales.

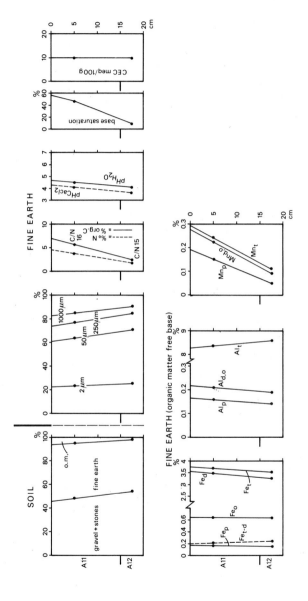

Figure 12   Depth functions of various characteristics of soil profile Haarts 2

37

These soils are well-drained and have a very thin O
horizon, covering an A1 horizon, the latter directly overlying the slightly weathered bedrock (profile Haarts 2).
Locally a very thin transitional horizon (= A12 horizon) is
present below the surface horizon. Because of the shallowness and weak development, the soils on the plateau tops
are considered as young soils; they are classified as Lithic
Udorthents or as Lithic Umbric Dystrochrepts (Soil
Taxonomy 1975) and as Dystric Regosols or as Rankers
(FAO Unesco 1974) (see also A II 6.5). Similar soils have
frequently been described in Western Germany and are
developed on small plateau summits and steep slopes directly
on shales and graywackes. These soils are classified as
Braunerde-Ranker, which represents a transitional stage to
the development of an "Oligotrophe Braunerde" (Dystrochrepts)
(Mückenhausen 1976).

Profile descriptions are given in appendix II and some
characteristics of these soil profiles are shown in fig. 12
and all colours mentioned in the text are moist. The
surface horizon is dark brown (7.5-10 YR 3/2), rich in
organic material (org. C 5.6 %) of an acid mull-type (C/N
ratio 15.6). This horizon has a high permeability. The
thin "Cambic-like" A12 horizon has a brown to dark brown
(7.5-10 YR 4/4) colour, a C/N ratio of 14.8 and shows a
clear and smooth boundary to the slightly disintegrated shales.
These soils have a gravelly-stony loamy texture; their
water holding capacity seems to be rather low due to their
shallowness. The soils are distinctly acid (pH_{H_2O} 4.5 to
4.1, pH_{CaCl_2} 4.1 to 3.6), but they have a

medium cation exchange capacity (ca. 10 m eq/100g dry soil)
because the A11 horizon (45 %) is much higher than that of
the A12 horizon (8 %) (fig. 12), because especially calcium
and to a lesser extent magnesium and potassium are enriched
by litter fall. This phenomenon is also reflected by the
somewhat higher pH (pH_{H_2O} = 4.5) in the surface horizon and

might be attributed to "insoluble humin" which is a
characteristic of the acid mull-type of humus (Duchaufour
1976).

These soils represent at this moment a rather early
stage of weathering and soil formation. This could be concluded from the shallowness of the solum and the slight
development of a (very thin) A12 horizon. It is also indicated by the uniformity in grain-size distribution throughout
the solum and the only slight decrease of gravel and stones
in the surface horizon. Clear weathering and soil forming
phenomena can be detected with the help of depth functions
of the pedogenic oxides. Explanations or descriptions of
the various extracting methods, eg. dithionite-citrate-
bicarbonate, ammonium-oxalate and the sodium pyrophosphate,
for the secondary iron, aluminium and manganese forms in soils
are given in appendix II. Although the results have to be
treated with care to regard to homogeneity of the parent
materials and the influence of the biocycle (Blume &
Schwertmann 1969), it is quite clear that the depth functions
of these soils (see fig. 12) also indicate only an initial of
the high organic matter content. The base saturation of

stage of soil formation and weathering. Only a small increase of dithionite-extractable iron (Fe_d; free iron) could be observed in the A11 horizon, indicating that very little pedogenic iron oxides have formed in this surface horizon compared to the A12 horizon. This conclusion is also confirmed by the uniform distribution of oxalate-extractable iron (Fe_o; amorphous iron) throughout the solum, the uniform low Fe_o/Fe_d ratio and the only very slight decrease of silicate-bound iron (Fe_{t-d}; because no magnetite, carbonates or only very small traces of ilmenite are present) towards the surface. The uniform distribution of the "free aluminium oxides", indicated by the Al_d or Al_o (no gibbsite present) contents suggest the same lack of strong weathering. The increase in the "free manganese oxides" (Mn_d or Mn_o) towards the surface can be attributed only to accumulation of organic material, since a high proportion of the total manganese participates in the bio-cycle. This is corroborated by the increase of the pyro-phosphate extractable manganese (Mn_p) content towards the surface. The main changes in chemical and mineralogical composition have occurred in the formation of soil (material) from the bedrock, but this will be discussed in Part B 1.

6.3   Soils on the valley slopes

6.3.1 Introduction

On the valley slopes polygenetic soils have developed in slope deposits. Three groups of soils have been recognized in the two geomorphic units. On the gentle upper slopes soils are found with an (O)-A1-A3-B2-IIB2tb-IIC profile. On the upper part of the steep valley slopes on many sites soils with an (O)-A1-B1-B2-IIC-R1 profile occur, but also soils with an (O)-A1-B1-B2-IIB2tb-IIC-R1 profile. On the steep lower valley slopes soils with an (O)-A1-B1-B2-B3(+R1)-IIB2tb-IIC+R1-IIIR1+C profile could be recognized. A schematic cross-section showing the distri-bution of the soils is represented in fig. 11.

6.3.2 Parent material

At least two slope deposits could be established on the valley slopes. The older deposit, although not yet definitely dated, probably is of Würm age. This deposit consists of relatively unweathered hard shales and very few quartzitic sandstone fragments of gravel and stone size, embedded in a loamy matrix. Micromorphological investigations give the same impression of the character of the material. Embedded in the S-matrix are sharply bound lithorelicts, showing no weathering rims or zones (Kwaad & Mücher 1977). It is assumed that this mixture of unweathered rock fragments and loam represents a mass-wasting deposit, considering the large size of rock fragments (up to 20 cm

diameter) and the overall chaotic appearance of the deposit.
This older deposit contains neither charcoal fragments, nor,
except in the upper few cm's, pollen (see appendix II,
table 59), while an argillic horizon with yellow-brown strong
continuous ferriargillans has developed in the deposit.
The presence of a few strong continuous red-brown papules
in this layer corroborates that the material has been transported. The greatest thickness of this slope deposit was
found to be on the steep lower slopes (profile Haarts 1).
Here erosion was apparently less dominant in the Middle
Ages, which might be deduced from the thickness of the
deposit and the argillic horizon. On the upper parts of
the steep valley slopes this older deposit is rather thin
and has perhaps been partly removed by erosion, which might
also be indicated by the lack of the argillic horizon on
some sites ((O)-A1-B2-IIC profile; profile Nothum 3). On
the gentle upper slopes the deposit is also thin, but it
does contain a thin argillic horizon (((O)-A1-A3-B2-IIBtb-
IIC profiles; profile Nothum 1).

On the steep lower slopes beneath this slope deposit
another probably older mass-wasting deposit having a
grèze-litée character occurs. This lower saprolite consists
of gravel and stones (81 %), a fine earth fraction, which
is characterised by a lower silt and clay content and a
higher coarse sand fraction (III?R1+C) horizon of profile
Haarts 1). From this presence of this grèze-litée like
deposit it may be concluded that the overlying saprolite
with the argillic horizon is another slope deposit and not
a strictly *in situ* weathering product of the bedrock.

A young (colluvial) slope deposit overlying the older
one (with the argillic horizon) has been found all over the
valley slopes. More or less the same variation in thickness
in this younger deposit as in the older one could be
established. On the gentle upper slopes and the upper parts
of the steep lower slopes it has a thickness of about 40 to
50 cm, but on the steep valley slopes a thickness of 80 to
90 cm occurs. The thickness of this deposit is of great
importance for some characteristics and the classification
of the truncated subsoil with the argillic horizon, according
to the Soil Taxonomy (1975) (see A II 6.5). This slope
deposit is characterised by 1) a young Subatlantic pollen
spectrum with a clear zonation (zone X a, b; Firbas 1952);
2) many (large) charcoal fragments; 3) fresh greyish rock
fragments; 4) presence of pedorelicts in the form of some
yellow-brown papules and a few para-aggregates; 5) sharply
bounded and (sub-) rounded ferric nodules next to diffuse
ferric nodules, the latter apparently formed *in situ* (Kwaad
& Mücher 1977); 6) presence of a cambic horizon; 7)
gravelly and stony loamy texture, although its gravel and
stone content might be influenced by weathering and soil
formation (fig. 14). Kwaad & Mücher (1977) consider this
young slope deposit as a colluvial deposit, which was
originally rich in organic matter and which contained fecal
pellets. After deposition, mineralisation of the organic
matter is assumed to have occurred. That part of the slope
deposit which has lost much of its organic material now
contains the cambic horizon. Kwaad & Mücher considered that

the transitional A3 horizon of profile Nothum 1 still contains its organic matter, while the A1 horizon has been enriched in organic material by soil formation *in situ*, indicating that now soil formation is more important than colluviation.

6.3.3 The soils on the gentle upper slopes

The soils on the gentle upper slopes in the Haarts catchment are less shallow than the soils on the plateau summit and have a polygenetic origin. They have an (O)-A1-A3-B2-IIB2tb-IIC profile (profile Nothum 1) and are developed in the two slope deposits. These soils have been classified as Thapto-hapludultic Umbric Dystrochrepts. They represent a soil with a dark surface horizon (umbric epipedon < 25 cm thick) and a cambic horizon (Umbric Dystrochrepts) in a young colluvium deposited between 1450-1800 AD. This "surface" soil lies on a truncated soil profile with a remnant of an argillic horizon developed in the older slope deposit, which became more acid probably in a later stage of soil formation after deposition of the colluvium (see also A II 6.5).

These soils are well-drained with a very high biological activity (see micromorphological description) and have only a very limited O horizon (0-1 cm thick) which covers a rather thick A horizon. A representative profile description is given in appendix II and some characteristics of these soils are shown in fig. 13. The A1 horizon is very dark brown (10 YR 2/2), rich in organic material (org. C 4.65 %), of an acid mull type (C/N ratio 13). This surface horizon is distinctly acid (pH_{H_2O} 4.1, pH_{CaCl_2} 3.9) and has a gravelly silt loam texture, a moderate to weak medium crumb structure and a very high infiltration capacity due to its high porosity. Beneath the surface horizon an A3 horizon exists, with a dark brown (7.5 YR 3.5/3) colour, a gravelly silt loam texture, a weak fine crumb structure and a C/N ratio of 11. It is also distinctly acid (pH_{H_2O} 4.3, pH_{CaCl_2} 3.9). This transitional horizon possibly reflects, with respect to its organic carbon content and its colour, the original characteristics of the colluvial deposit (Kwaad & Mücher 1977). The cambic B horizon of the "surface" soil has a yellowish brown (10 YR 5/5) colour, a gravelly silt loam texture, a moderate coarse subangular blocky structure and a C/N ratio of 8. The decrease in the C/N ratio below 10 in the subsurface horizons is characteristic for non-cultivated soils in the (temperate) humid climates with a rather high biological activity and can be explained by the fixation of non-exchangeable NH_4^+ in these mineral horizons (Stevenson *et al* 1958). The base saturation status of the "surface" soil is low, ranging from 9 % in the A1 horizon to about 5 % in the cambic B horizon (fig. 13). Aluminium is the dominant cation at the adsorption complex of the "surface" soil (appendix II, table 43).

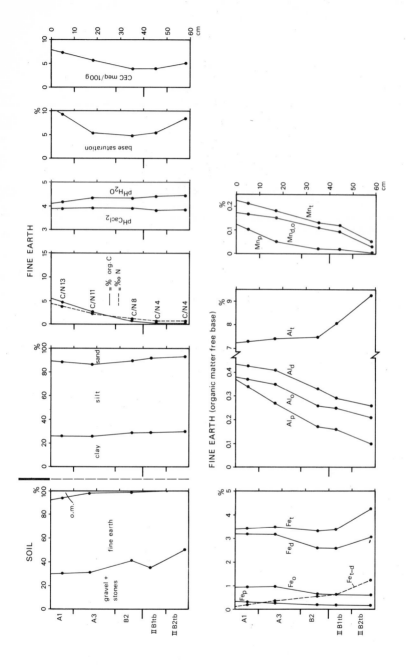

Figure 13   Depth functions of various characteristics of soil profile Nothum 1.

42

The "surface" soil belongs to the "Acid Brown Forest soils without podzolization and illuviation characteristics" (Duchafour 1977). This can be concluded from investigations of the thin sections (app. II, table 41) and the low Fe_p/Fe_d or Fe_p/Fe_o values throughout the profile, the last two ratios showing a tendency to increase towards the surface. The Fe_p/Fe_d ratio of 0.07 in the B horizon is typical of a cambic horizon (Duchafour 1976) indicating the very small influence of the metalorganic complexes on the behaviour of free iron. Also the free Al $(Al_d, Al_o)/$ clay ratio ranging from 0.02 in the A1 horizon to 0.01 in the cambic horizon suggests that degradation of the clay minerals is in a rather initial stage. The depth functions (fig. 13) in this "surface" soil for the various free iron compounds lead to a similar conclusion. They show an increase in the Fe_d, Fe_o and Fe_p contents going from the cambic horizon towards the surface horizon. No downward movement of pedogenic iron could be established. The opposite trend of the silicate bound iron (Fe_{t-d}) (fig. 13) indicates that the pedogenic iron oxides have formed from this silicate iron. Also the "activity ratio" Fe_o/Fe_d presents a picture which is normal for Dystrochrepts. The topsoil Fe_o maxima coincide with relatively high Fe_o/Fe_d ratios indicating a low degree of aging (crystallization) of the free iron compounds. This may be caused by unfavourable conditions for crystallization due to the retarding effect of organic material (Schwertmann 1966; Schwertmann *et al* 1968; Schwertmann 1969). Also the depth functions of the free aluminium and free manganese compounds give similar results (fig. 13). Apparently, although the environment is distinctly acid, the time of pedogenesis has been too short for the migration of free manganese oxides and a slight migration of free aluminium from the A1 horizon towards the A3 or the cambic horizon (often established for Dystrochrepts in Western Germany, Blume & Schwertmann(1969)) to be effective. From figure 13 it is also clear that the increase in the pedogenic manganese oxides towards the surface can be completely attributed to the increase in the pyrophosphate extractable manganese (Mn_p) content, indicating that the accumulation which has a relatively high manganese level (Duchaufour 1977) was and still is effective near the surface. The results support the investigations in the Vosges (France) for similar soils showing that the humification, resulting in the acid mull type humus, is very effective if the parent rocks contain more than 4.5 % free Fe_2O_3 (> 3.2 % Fe) (Duchafour & Souchier 1978). The slightly disintegrated parent rocks in the Haarts catchment fulfil this requirement.

The lower part of the soil profile, the "subsurface" soil which is developed in the older slope deposit, consists of an argillic horizon. This argillic horizon can be divided into two parts. A IIB1tb and a IIB2tb horizon, both consisting of a light grey to hellow (2.5 YR 7/3) and strong brown (7.5 YR 5/6) mottled gravelley silt loam with a low C/N ratio and a distinctly acid character (pH_{H_2O} = 4.4; pH_{CaCl_2} = 3.8). The subdivision has been made with the help of micromorphological investigations. The IIB2tb horizon contains many yellow and brown continuous ferriargillans, with some dirty or speckled brown continuous ferriargillans. The IIB1tb horizon has much less illuviation phenomena. These illuviation features consist of some yellow and brown strong continuous ferriargillans, more dirty or speckled brown continuous ferriargillans and in the upper part of this subhorizon also some matrans (see appendix II Micromorphological descriptions). The base status of this "subsurface" soil is also low, indicating the acidification after the formation of the argillic horizon and the deposition of the colluvial layer. This conclusion is confirmed by the high Al content at the adsorption complex (appendix II, table 43) indicating that clay migration is today inhibited by aggregation through aluminization of the clay (Schwertmann 1969) (see also A II 6.3.4). The depth functions of the "subsurface" soil give similar trends which have been established in Alfisols with a low base saturation, namely a constant Fe_d/clay content and no important degradation of the clay minerals, indicated by a rather constant free Al/clay content. Only the decrease in the total aluminium content going from the IIB2tb towards the IIb1tb horizon does not fit in the outline of the presented soil formation and remains unexplained.

6.3.4 Soils on the steep lower slopes

The soils in this geomorphic unit in the Haarts watershed also are polygenetic and have developed in the two slope deposits. Two types of soil could be distinguished in this geomorphic unit. On some places in the upper convex part of the steep lower valley slopes soils with an (O)-A1-B2-IIC profile occur, which have been classified as Umbric Dystrochrepts. Nevertheless, on the greater part of the upper convex area of this geomorphic unit and everywhere on the steeper lower parts of the valley slopes, soils with an (O)-A1-B-IIB2tb-IIC+R1-III?R1+C horizon exist. These soils have been classified as Thapto-hapludalific Umbric Dystrochrepts.

The Umbric Dystrochrepts (see appendix II, profile Nothum 3) on the upper convex parts are closely related to the soils on the gentle upper slopes. The "surface" soil (A1 + B1 + cambic B horizon) is quite similar concerning the macroscopic features of colour, texture, structure and consistency. Also the chemical data such as base saturation, CEC, composition at the adsorption complex, C/N ratio are more or less the same. The lower part of the IIC(+R1) horizon, probably represents the older slope deposit, although this has not yet ultimately been established, due to the lack of micromorphological data from these horizons and the absence of the grèze-litée-like material, which is present on the steeper lower part in this

geomorphic unit. However, due to the presence of soils
with a rather thin argillic horizon and the characteristics
of the saprolite similar to those of the older slope
deposits on similar places in these upper convex areas, it
is concluded that the lower part of the soil profile
represents a more truncated phase of the soil in the older
slope deposits.

On the steep lower slopes soils with an (O)-A1-B(+R1)-
IIB2tb-IIC+R1-III?R1+C profile are widely distributed.
These soils have been classified as Umbric Dystrochrepts
developed in the young slope deposit, covering a remnant
of a Hapludalf in the older slope deposit (see appendix II,
profile Haarts 1). Also in the adjacent watershed in the
same geomorphic unit similar soils have been developed
(see appendix II, profile Scheissgrond 1). These soils
are well-drained and have a rather high permeability. The
hydraulic conductivity in the topsoil (A1 horizon) is more
than 15 m day^{-1}, while the permeability is about 5-6 m day^{-1}
for the B2 horizon, and 1-3 m day^{-1} for the B3 horizon.
The III?R1+C and IV?R1 horizons have an extremely high
permeability ($K \gg 50$ m day^{-1}). The hydraulic conductivity
was determined by the inversed auger hole (Porchet) method
(Kessler & Oosterbaan 1974). One should always keep in
mind that methods for measuring hydraulic conductivity in
soil layers above the water table may differ from those of
the saturated hydraulic conductivity due to swelling
properties of the soil. However, no significant changes in
structure have been observed for these soils up until now,
due to the high gravel and stone content and the mineralogica
composition.

The soil has been subdivided as in the previous
section into a "surface" soil (Umbric Dystroschrept) and
a "subsurface" soil (remnant of a Hapludalf). The "surface"
soil, at least the upper part till the B3+R1 horizon,
shows a high biological activity. This soil has a very
thin (0-1 cm) very dark brown (10 YR 2/2) O horizon, which
consists of partly decomposed organic matter. Beneath
this organic horizon a dark brown A1 horizon is found,
which could be divided into an A11 horizon (10 YR 3/3,
strong fine granular structure) with a clear boundary and an
A12 horizon (7.5 YR 3.5/3, strong fine sub-angular blocky
structure). This A1 horizon is rich in organic material
(org. C 7.2 %) of an acid mull type (C/N ratio 16),
distinctly avid (pH$_{H_2O}$ 4.0, pH$_{CaCl_2}$ 3.8) and has a gravelly
and stony loam texture (fig. 14 and appendix II). Below
this surface horizon, a cambic horizon occurs, with a
brown to strong brown (7.5 YR 5/5) colour, a gravelly and
stony loam texture, a moderate coarse subangular blocky
structure and a rather high C/N content of 14. This high
C/N content still reflects the characteristics of the
colluvium and the relatively young stage of this part of
the cambic horizon. A transitional B3 horizon has
developed which has been subdivided on its stone content,
colour and development of pedogenic structure into a B3 and
B3+R1 horizon. The B3 horizon has a brown (7.5 YR 5/4)
colour, a gravelly and stony (52 %) silt loam texture, a

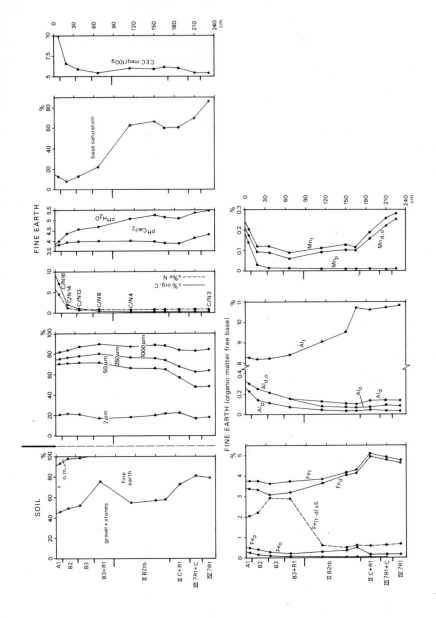

Figure 14   Depth functions of various characteristics of soil profile Haarts 1

46

weak coarse subanvular blocky structure and a C/N ratio
of 8. The B3+R1 horizon consists of a yellowish brown
(1O YR 5/8) very gravelly and stony (74.5 %) silt loam,
with a weak moderate subangular blocky structure and a low
C/N ratio. Thus, in this part of the cambic horizon the
normal decrease in the C/N ratio for non-cultivated soils,
with the exception of Spodosols, in the temperate humid
climate occurs. From palynological investigations it could
be established that the B2, B3 and B3+R1 horizons have
developed in the young slope deposits (presence of
Fagopyrum and many charcoal fragments, Slotboom, oral
communication) after 1800 AD (presence of *Picea*) under
forest conditions (relatively high arboreal/non-arboreal
pollen ratio: see appendix II, table 59). These B
horizons do not show any sign of illuviation features,
also indicating a formation of a cambic horizon in a non-
agricultural environment, and in accordance with the
composition at the adsorption complex (mainly Al) of the
A1, B2 and B3 horizons. In adjacent areas under arable
land illuviation features (matrans, dirty specked brown
ferriargillans) in (cambic) horizons in the young slope
deposits have been established (Kwaad & Mücher 1978).
Nevertheless, investigations of this sections show that
the fabric of the B2 and B3 horizons is somewhat
different. The B3 horizon is more massive than the B2
horizon. The latter consists completely of loose coprogene
material, possibly indicating a relatively slow
sedimentation environment in relation to the biological
activity, while the B3 horizon seems to reflect somewhat
more rapid depositional conditions. The base saturation
of the"surface"soil is rather low, ranging from 12.5 %
in the A1 horizon to 22 % in the B3+R1 horizon, and shows
the characteristic distribution, with a minimum value just
below the topsoil, for the Dystrochrepts (fig. 13)
(Duchafour 1977). This figure also indicates that the CEC
of the surface soil is greatly influenced by the organic
matter. The various depth functions for the pedogenic
iron, aluminium and manganese oxides presents the same
picture as those for the soils on the gentle upper slopes.
The highest values for the Fe_d, Fe_o, Fe_p, Al_d, Al_o and Al_p
compounds occur in the A1 horizon, decrease going towards
the B2 and B3 horizons. There has been no downward movement
of pedogenic oxides and the above demonstrates, together
with the lowest Fe_{t-d} values, that the A1 horizon is the
zone of maximum weathering. No podzolization features
could be established. This is confirmed by the low
Fe_p/Fe_d, low Fe_p/Fe_o and $(Fe_p + Al_p)/Fe_d$ ratios throughout
the surface soil, although there are slight increases
towards the surface. The iron bounded in the silicates
(Fe_{t-d}) shows a constant decrease towards the surface.

The opposite trend for the pedogenic iron oxides indicates
that these compounds have formed from this silicate iron.
The low "activity" ratio (Fe_o/Fe_d) for the pedogenic iron
oxides is also notable. The very low free Al/clay ratio
in this surface soil suggests that degradation of the clay
minerals is limited. The depth functions of the pedogenic

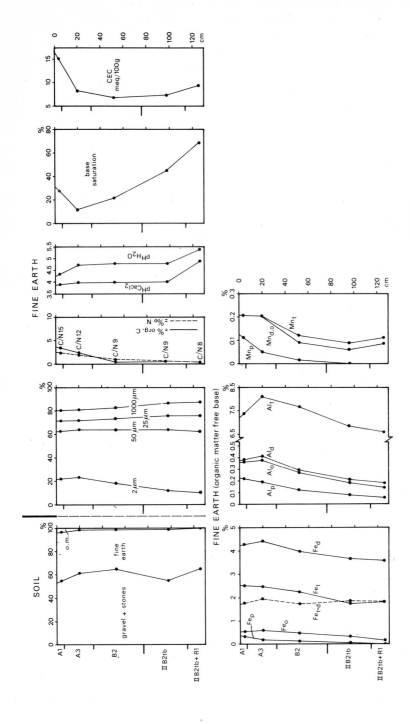

Figure 15   Depth functions of various characteristics of soil profile Scheissgrond 1

48

manganese compounds show similar trends. Here also a
small increase towards the surface, due to an increase in
the Mn content by accumulation of organic rich material in
the topsoil, occurs. Like the Dystrochrepts on the other
geomorphic units, manganese in the soils is present
purely as free manganese oxides.

The "subsurface" soil is a remnant of an Alfisol.
The IIB2tb horizon consists of a yellowish brown (10 YR
5/6) very gravelly and stony silt loam with a low C/N
ratio, a less acid character (pH_{H_2O} 5.1; pH_{CaCl_2} 3.9) and
a base saturation of about 60 percent, with magnesium as
the dominant cation at the adsorption complex. This
argillic horizon has many yellow brown strong continuous
ferriargillans, a few of which are deformed to papules.
In the upper part of this horizon most of these
ferriargillans are dirty or speckled brown weak continuous
ferriargillans and in the uppermost part of this argillic
horizon also a few matrans occur. These various illuviation
features represent the changing environmental conditions
during and/or after the formation of the argillic horizon.
The strong continuous yellow-brown ferriargillans are
considered to be the result of the formation of the argillic
horizon under natural conditions during the Early Holocene
(landscape stability). The matrans (agricutans) have to
be restricted to a period with agricultural practices
(arable land) after the formation of the argillic horizon,
because these features only occur in the uppermost part
of this horizon, but before deposition of the colluvial
slope deposit. The dirty or speckled brown weak continuous
ferriargillans possibly represent a period in between the two
mentioned before, with a more extensive human occupation
(see also Kwaad & Mücher 1978). The various depth
functions of Fe, Al and Mn provide some additional information. The difference in silicate bounded iron in the
IIB2tb and the "surface" soil is striking, indicating the
polygenetic character of the whole soil profile. Within
the subsurface soil the total Fe, Al and Mn levels decrease
going from the IIC+R1 horizons towards the IIB2tb horizon.
The parallel behaviour of the free iron (Fe_d) and free
manganese (Mn_d or Mn_o) compared with the total iron and
total manganese is obvious, indicating the very low silicate
bounded iron and manganese content in the "subsurface" soil.
Also no increases for the free Al could be established
going from the IIC+R1 to the IIB2tb horizon although the
total aluminium content decreases rather strongly. The
activity ratio (Fe_o/Fe_d) is relatively low for an argillic
(0.08-0.12) but is for the argillic horizon considerably
higher than in the IIC+R1 horizon. For additional data
see appendix II.

6.4 Soils in the dry upper valley

In the upper part of the Haarts valley soils have
developed in at least two slope deposits. These soils
have an (O)-Al-C-IIB2tb-IIBtb-R1 profile (see appendix II,
profile Haarts 3) and are present in the dry valley bottom.

The soils will be only briefly discussed because they do not give additional information on weathering phenomena.

These soils might be classified as Typic Haplumbrepts (Humic Cambisols) because they are recognised at first sight in the field as soils with am umbric epipedon, A1. (7.5 YR 3/2) - A3 (5 YR 3/4) horizons, on a genetically related cambic B (10 YR 5.5/6) horizon. However, from study of the soils in more detail in the field, and supported with micromorphological and palynological investigations, a quite different interpretation results (see appendix II and fig. 11). Here a slightly modified text from Kwaad (1977) and Kwaad & Mücher (1977) is presented. These authors studied these soils in detail in relation to the landscape history in the Birbaach catchment. The following points are given for a polygenetic development of these soils.

First, there is the thickness of the colluvium. On the side slopes of the dry valleys A horizons with a maximum depth of 5 to 9 cm occur, whereas here the organic matter containing horizon is 47 cm and in the Groendchen dry valley it is 70 cm, including the stone line. It seems only logical to assume that the valley side slopes constantly lose material, which accumulates in the dry valley centres.

A second point is that the top 47 cm contains significantly fewer (14 %) and smaller gravel and stones than the underlying horizons (55 %) and that this cannot be a consequence of a more intensive weathering of the upper part of the soil, as the rock fragments are hard without any sign of a weathered outer zone.

The presence of a stone line at the base of the colluvium is a third indication for a depositionary nature of the overlying material.

Fourthly, there is the presence of charcoal in the colluvium which is not surprising in view of the widespread charcoal-burning practices, until 1840, in the woods of northern Luxembourg for the iron smelting works in the south (Steffes 1965) and the forest burning for agricultural purposes (Nepper 1904).

From the microscopic examination of this sections the colluvium appeared to overly an argillic horizon with well developed illuviation ferriargillans. No A2 horizon was found between the colluvium and the B horizon, which therefore was designated as a IIBtb horizon.

Finally, there is micromorphological and palynological evidence that the bulk of the colluvium is not subjected to reworking by the soil microfauna, which, if it was, would make it an A horizon *in situ*. The irregular topography of the lower boundary of the colluvium apparently is due to animal burrowing. The burrows, however, are large and isolated, not affecting the bulk of the horizon. The pollen in two burrows appeared to be derived directly from the present litter layer and to be dissimilar to the pollen in the surrounding soil material. The welded character of the fecal pellets below 15 cm and the general lack of channels and pedotubules also do not point to an important biologic activity in the lower part of the colluvium.

So it can be concluded that in the dry valley bottom of the Haarts catchment <u>soils</u> have developed in at least two slope deposits and that these soils have to be considered as a (remnant of) buried argillic horizon covered by a colluvium with an ochric epipedon. Therefore these soils have to be classified as Thapto-hapludalfic Udorthents or as Mollic Hapludalfs (Dystric regosol, on a buried luvisol), although few data on the base saturation of the buried argillic horizon are available (see also A.II.6.5).

The older slope deposit with the argillic horizon is similar to the older slope deposits with an argillic horizon on the valley slopes. These deposits represent a periglacial (Würm?) pre-Allerød mass-wasting phenomena. More can be said on the origin of the younger colluvial slope deposit. This colluvium has a slight gravelly-stony loam texture and can be divided with the help of palynological data (appendix II, fig. 35), into two parts. An older, lower part (10-47 cm below the surface) is characterised by a high non-tree pollen/tree pollen ratio. Apparently this older colluvium has not been formed in a woodland environment, which seems to have been confirmed by an old land use map (1:25,000) of Comte Ferraris (1777) and the presence of matri-(ferri)argillans in the upper part of the buried argillic horizon. An important landmark is the presence of buckwheat *(Fagopyrum,* introduced about 1460 AD), the curve of which starts at a depth of 37 cm. In the upper, younger part (0-10 cm) of the colluvium spruce *(Picea)* is found, which was introduced in this area about 1800 AD by reafforestation. Also the high percentage of tree pollen indicates that this upper part was formed under the present coppiced forest and splash erosion is (almost) completely responsible for the formation of this younger colluvium (Imeson & Kwaad 1976; Kwaad 1977).

6.5   Some remarks on the classification of the soils

In this section some comments will be made on the classification of the soils according to the Soil Taxonomy (1975) and the FAO/UNESCO Soil Map of the World Legend (1974). The soils on the small plateau tops, which are monogenetic, give no difficulty in this respect (appendix II; profile Haarts 2). However, the soils on the other geomorphic units in the catchment pose some problems because they are polygenetic soils developed in at least two parent materials. For this reason, a short review is given of those aspects of the genesis of these soils, which are relevant for their classification.

As shown in the previous paragraphs, the soils on the valley slopes and in the dry valley have developed in at least two parent materials. An older slope deposit, with an argillic horizon, not present on the upper convex parts of the steep lower slopes, and a younger slope deposit, with a cambic horizon on the valley slopes, and only an ochric epipedon in the upper dry valley. So, from a geogenetic point of view all of these soils are closely related to each other. Moreover, it seems that the thickness of the slope deposits, which is closely related to the various geomorphic units, is a dominant factor for the classification of the soils. The effectiveness of the erosion on the older slope deposits on the upper convex part of the steep lower slopes, after the formation of the argillic horizon, seems to be evident because this horizon probably has been removed, so these soils should be classified as Umbric Dystrochrepts (appendix II; profile Nothum 3). Erosion was less effective on the other geomorphic units of the valley slopes, resulting in a thin older slope deposit with a very thin argillic horizon on the upper slopes, and a thicker older slope deposit with a relatively thick argillic horizon on the steep lower valley slopes and dry valley (table 8). A younger slope deposit accumulated between about 1460 and 1800 AD. With regard to its thickness, the younger slope deposit on the valley slopes shows the same distribution as the older slope deposit: a rather thin layer on the gentle upper slope and the upper convex part of the steep lower slope (40-50 cm) and a thicker layer on the steep lower slope (85-95 cm). Moreover, a thin younger slope deposit was formed in the upper dry valley (45-55 cm). In this younger slope deposit on the valley slopes a Dystro-chrept (Dystric Cambisol) showing marked acidification has developed, but at those places where the younger slope deposit was thin, the acidification also affected (the upper part of) the argillic horizon of the older slope deposit. This has resulted in a strong lowering of the base saturation of this part of thy soil profile. Therefore, the buried part of the soil profile on the gentle upper slope has to be classified as a Hapludult (Acrisol). This Hapludult is considered to be buried (according to the Soil Taxonomy 1975) because it has a surface mantle between 30 and 50 cm thick and the thickness of that mantle is at least half that of the preserved argillic horizon. There-fore these soils have to be classified as Thapto-hapludultic Umbric Dystrochrepts (appendix II; profile Nothum 1). The buried part of the soil profile in the dry valley also has to be classified according to the FAO/UNESCO system as an Acrisol, but according to the Soil Taxonomy as a Hapludalf, as only the upper part till 90 cm of the argillic horizon has a base saturation less than 35 % (appendix II; profile Haarts 3). On the steep lower part of the valley slopes, due to the greater thickness of the younger slope deposit, acidification resulting in a low base saturation has not (yet) influenced the argillic horizon and therefore the buried parts of these soils have to be classified as Hapludalfs (Luvisols).

Although the truncated soils on the valley slopes are buried by younger materials, these materials are definitely not unaltered as they have a cambic horizon and therefore do not satisfy the requirements for the buried soils (Soil Taxonomy 1975) (appendix II; profile Haarts 1). Nevertheless, all of these soils were classified for the Soil Taxonomy system with the thapto prefix. Only the soils with an argillic horizon in the dry valley are buried

with largely unaltered colluvium. These soils fulfil the
requirements of buried soils if the colluvium is more than
50 cm thick. For the dry valley two types of soils with
regard to the soil classification thus have to be
distinguished: 1) a Mollic Hapludalf (Orthic Acrisol) for
the soils with a surface mantle less than 50 cm thick. In
the author's opinion, however, it seems preferable to
classify these soils as Umbric Hapludalfs, although these
are not represented inthe Soil Taxonomy (1975), because
this gives a better expression of the low base saturation
of the top soil; 2) a Thapto-hapludalfic Udorthent
(Dystric Fluvisol on a buried Acrisol) for those soils
with a surface mantle of more than 50 cm.

WATER-ROCK INTERACTIONS (WEATHERING)

In this part the alteration of the (very) low-grade
metamorphic shales will be discussed. Rock at or near the
earth surface is attached by both physical and chemical
processes, which may transform the shales to such an
extent that the reactant (weathering) products differ
considerably from the original materials. Breakdown of
rock at the earth surface results from a direct contact with
the atmosphere (weather), therefore the term "weathering"
is used to indicate these transformation processes.

Although physical (mechanical) weathering is
indispensible for an effictive operation of the chemical
weathering, its influence on the disintegration of the rocks
in the Haarts catchment has not been sufficiently variable
to give rise to different paths of chemical weathering.
Only the latter, the water-rock interactions, have been
taken into account in this part. Chemical weathering is a
process by which atmospheric, hydrospheric and biological
agencies act upon and react with the mineral constituents
of rock within the zone of influence of the atmosphere,
thereby producing relatively more stable, new mineral
phases (Reiche 1950).

In order to get a complete picture of the chemical
weathering process one has to approach the subject from
several angles. Firstly the chemical and mineralogical
composition of the (parent) rock has to be known, to define
the initial composition. Secondly, the chemical and
mineralogical composition of the weathering materials and
the soil horizons in the saprolite have to be traced in
order to establish the various stages of the weathering
sequence(s) with their relevant reactant products. Thirdly,
because chemical weathering can be considered as a reaction
of a mineral assemblage with an aqueous phase, which is not
in (overall) equilibrium with this mineral assemblage, the
chemical composition of the initial aqueous solution
(atmospheric precipitation) has to be known. The changes
in the chemical composition of the aqueous solution, when
precipitation percolates through the soil and weathering
materials until it leaves these as spring waters, have also
to be investigated in order to obtain information on the
transformation of the mineral assemblage in these environments. So it can be concluded that an optimal insight in
the chemical weathering process can only be obtained if a
simultaneous investigation has been carried out on the
chemical and mineralogical composition of the solid phases
(parent rocks and reactant products) and the chemistry of
the precipitation-, soil- and springwaters. All of these
aspects will be discussed in this part, but also a
theoretical model of the parent rocks, with an aqueous
solution in equilibrium with relevant partial CO_2 pressures
will be presented and compared with the mineralogy of the
weathering products and the chemical composition of the soil-
and springwaters. Finally the influence of other relevant

physiographic factors, like landscape history (erosion and depositional features), hydrology, vegetation, soils and anthropogenic activities, which are already discussed in part A, will be incorporated into this part in order to emphasise their importance for the mineralogical composition of the weathering material.

1. CHEMICAL AND MINERALOGICAL COMPOSITION OF THE SOLID PHASES

1.1 Rocks

1.1.1 Introduction

The rocks in the Haarts catchment are marine sediments in origin and form part of a Hercynian massif, which includes the Ardennes and the Rheinische Schiefergebirge. The original sediments have been strongly influenced by diagenetic and/or low-grade metamorphic processes. The degree of diagenesis and/or metamorphisms of these rocks varies within short distances in the Hercynian massif (Weber 1972). The mineralogical composition of the originally fine-grained pelitic rocks is closely related to the degree of diagenesis and low-grade metamorphism, and therefore the relevant (clay) mineralogical transformations by these processes will be discussed briefly. For more detailed information the reader is referred to Weaver 1956, 1958, 1960; Kubler 1964, 1967; Turner 1968; Dunoyer de Segonzac 1969; Weaver & Beck 1971; Winkler 1974, and Velde 1977. Although recently knowledge about high-grade diagenetic shales and low-grade metamorphic schists has increased considerably, the evolution of clay minerals is still too little understood to give precise limits for the various stages in the diagenetic evolution and (very) low-grade metamorphism (Velde 1977). The following provisional scheme of Dunoyer de Segonzac (1970) has been used:

1. Early diagenesis; in this stage all of the detrital clay minerals are (meta) stable; some of them undergo aggradation of Mg, K and Na (various mixed-layers) and also new formation of clay minerals (montmorillonites) occurs.

2. Middle diagenesis; in this stage the sediments are strongly compacted and lose at least 50% of their pore waters, but they remain porous. In acidic siliceous environments depending on different thermal conditions, kaolinite and dickite are formed by neoformation, whereas in alkaline environments, rich in potassium and magnesium, illite and chlorite are formed via mixed-layer intermediates. The illitization process of montmorillonites requires also a certain degree of compaction to expel one of the two interlayer water sheets.

55

3. Deep or late diagenesis; in this stage the temperature is higher than $100^{\circ}C$, pressure increases and porosity becomes very low. The transformation of clay minerals by aggradation is very effective and montmorillonites are illitized or chloritized by dehydration and absorption of potassium and magnesium. Also, the irregular "mixed-layers" are replaced by regular ones such as allevardite or corrensite. Kaolinite becomes unstable and is transformed to dickite in the acidic environment and destroyed in the alkaline environments. The crystallinity of illite increases and a polymorphic transformation 1Md $\rightarrow$ 2M$_1$ takes place. In this stage the assemblage of the clay minerals becomes uniform, independent of the initial lithology, producing "illite and chlorite facies".

4. The anchizone; this is the transitional zone to metamorphism. Illite and chlorite are almost the only phyllosilicates and the mineralogy of clays is completely independent of the initial lithology. However, dickite as well as pyrophyllite associated with allevardite can be observed. The crystallinity of illite defines the limits of this zone (Weaver 1960; Kubler 1964), which coincides more or less with the very low-grade metamorphism stage of Winkler (1974). It is generally considered that the formation of metamorphic schistosity begins in the anchizone (Kubler 1967).

5. The metamorphic epizone; the clay material (illites and chlorites) loses gradually its "clay" properties across the anchizone by increasing its particle size while lattice-re-ordering continues. In this zone, which is equivalent to the low-grade metamorphic stage of Winkler (1974) and the green schist facies (Turner 1968), phyllites with "sericite" and chlorite are produced. The illites of the deep diagenetic and anchizone are replaced by white micas. The latter often differ from real muscovites by a tetrahedral charge less than 0.9 (Si:Al > 3:1) and an increased octahedral charge due to substitution of iron and magnesium for aluminium. These micas are called phengites, although the sodium equivalents, such as parogonitic muscovites and paragonites also occur.

The Lower Devonian rocks in the Oesling belong to the anchizone to low-grade metamorphic region, therefore some general remarks can be made about the expected clay-mineralogical composition. Montmorillonites and kaolinite should not be present because both minerals will have been destroyed already in the deep diagenetic stage. This should be the case also for irregular mixed-layers, but regularly ordered mixed-layers such as corrensite and allevardite sometimes have been found in the mineral assemblage of the anchizone up to the border of the low-grade metamorphic zone. Chlorite and micaceous minerals are expected to be the dominant phyllosilicates in the rocks.

The major reaction common to most pelitic
assemblages within this range is illite → mica + chlorite
(Winkler 1964) or illite + chlorite → mica + chlorite,
and the 1Md "mica" or illite becomes a $2M_1$ polymorph
(Velde 1977). However, within this high-grade diagenetic/
low-grade metamorphic range many of the micaceous minerals
(10Å) contain chlorite layers which can be established
by the asymmetry of the *ool* illite peaks (the low angle
side of the 10 Å peak and the high angle side of the 5 Å
peak (Weaver & Beck 1971). Also the 14 Å and 4.75 Å peaks
of the chlorites often show an asymmetry towards the 10 Å
and 5 Å illite peaks, suggesting that some illite is
interlayered with the chlorite component. If it is true
that much of the compositional difference between illite
or phengite and muscovite is due to the presence of interlayered chlorite, then it is likely that the "chlorite"
interlayering is still present in low-grade metamorphic
micas (Weaver & Beck 1971). At a somewhat higher grade of
metamorphism muscovite and chlorite minerals are formed.
From bulk chemical analysis it seems that there is no
systematic increase in potassium nor a decrease in sodium
content of argillaceous rocks upon increasing temperature
and pressure. Albitization, representing a redistribution
of sodium within these rocks is a common phenomenon where
mixed-layered minerals are predominant in clay assemblages;
it is especially evident in the illite-chlorite zone
(Velde 1977). In the following sections the expected
mineral composition will be compared with the actual
situation in the Haarts catchment.

1.1.2 X-ray analysis

X-ray analysis of the mineral assemblage in 34
powdered rock samples of rocks underlying the Haarts
catchment have revealed the presence of K-mica (illite),
quartz, albite, chlorite, hematite and rutile, and, in the
rocks with a relatively high CO_2 content, also siderite
(appendix II, table 70). X-ray diffractograms were made
for the fraction < 2 μm of the powdered rock and indicate
that K-mica (illite) (10 Å), chlorite (14 Å) and a mixed
layer illite-chlorite (24 + 12 Å; no shift with glycerol,
no shift upon heating to 550°C) have been established
(appendix II; fig. 31c). Illite and chlorite are more
abundant in the more fine-grained part of the rock, while
quartz and albite predominate in the coarser part. Illite
is the most abundant clay mineral. Chlorite and the mixed-
layer illite-chlorite are present in only small amounts;
no kaolinite or expandable irregular mixed-layers could be
observed.

The order of the illite lattice or illite crystallinity
can be used to determine the relative degree of high-grade
diagenesis or low-grade metamorphism. This ordering is
mainly a function of the temperature (Kubler 1966). The
sharpness ratio (Weaver 1961) was used to determine the
degree of illite crystallinity. It is the ratio of the
height of the illite 001 peak at 10 Å to the height above
the base of 10.5 Å. For all rocks the sharpness ratio varies
between 3.5 - 5.5, indicating the very weak to weak meta-

morphism of the transitional anchizone. The 10 Å peak is somewhat asymmetrical to the low angle side and the 5 Å peak towards the high angle side, suggesting that a chlorite layer is interlayered in the K-mica structure (phengitic composition) (Weaver & Beck 1971).

The data provided by the X-ray analysis have been used for the epinorm calculations in the following section.

1.1.3 Chemical data and normative mineralogical composition

Mainly due to the differences in their phyllites or quartzo-phyllites habitus the rocks in the Haarts catchment show only some variation concerning their chemical and mineralogical composition. The low-grade metamorphic shales (phyllites) consist in general, according to Niggli's epinorm calculations (Burri 1964), or 30-45 % quartz, 30-50 % muscovite, 8-12 % chlorite, 5-8 % albite and minor amounts of hematite, rutile, apatite and traces of pyrite and siderite (see appendix II, table 69). The normative mineralogical composition of the low-grade metamorphic "sandy" shales (quartzophyllites) consist of 55-70 % quartz, 12-20 % muscovite, 4-11 % albite, 3-7 % chlorite, with also minor amounts of hematite, rutile and apatite and traces of siderite and pyrite. Both representatives contain a very low CO_2 content. However, the "sandy" shales also include rocks with a relatively high CO_2 content (3.5 - 4 %) suggesting an appreciable amount of siderite, a fact confirmed by X-ray analysis (appendix II, table 70).

The results given by the epinorm calculations can of course be only comparative, because amongst other things the norm calculations were made with ideal (end) members and these will certainly not be found in the high-grade diagenetic/low-grade metamorphic range. However, some remarks can be made on the results of the various normative mineralogical compositions in relation to the amounts of the minerals present. The K-micas were calculated as an ideal muscovite, but are in fact sericites of more phengitic composition (see also B 1.12). This is corroborated by a strong positive correlation between the K_2O and the MgO content of the rocks (r = 0.92) (fig. 16) and a somewhat less strong correlation between the K_2O and the MgO+FeO content (r = 0.72), indicating that a part of the magnesium and probably some ferrous iron is built into the K-mica lattice. These phengites are also characterized by a Si/Al ratio greater than 3 in the tetraeders. Therefore it can be concluded that the results of the epinorm calculations are a little too high for quartz and chlorite and too low for K-mica.

All of the rocks in the area investigated contain little calcium. From this and the derived epinorm calculations it can be concluded that the plagioclases are represented by a pure albite (Ab_{100}). This would be characteristic for low-grade metamorphic pelites of the green schist facies (Turner 1968). Therefore the percentages of the normative albite seems to be realistic. It is also clear (appendix II, table 69) that the minerals of the

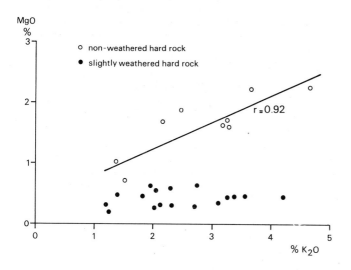

<u>Figure 16</u>   Linear regression between K_2O and MgO contents
of the (very) low-grade metamorphic rocks

epidote group like zoisite and clinozoisite, which are
diagnostic for the low-grade metamorphic stage (Winkler
1974) are absent or present at most as traces and not
detectable by X-ray analysis.

In the rocks with a low CO_2 content and in the
normative chlorite composition, the FeII/Mg ratio varies
between 0.78 and 0.85, suggesting ferrous iron-rich
chlorites which are often described in low-grade metamorphic
pelites (Turner 1968; Muller & Saxena 1977) (see also
X-ray analyses). The rocks with a relatively high CO_2
content have a somewhat higher ferrous iron content. In
the epinorm calculations all of the ferrous iron and
manganese was used for the formation of siderite, and even
a small part of the magnesium had to be used to consume all
of the CO_2 determined. Both Mn and Mg usually substitute
Fe^{II} in siderite, and some indications have been found that
a complete solid solution between siderite and rhodo-
chrosite and between siderite and magnesite exists
(Deer, Howie & Zussman 1966). In low-grade metamorphic
rocks siderites with a considerable amount of magnesium and
manganese have been established (Butler 1969). So it
would seem to be justified to conclude that the chlorite
in the rocks with the high CO_2 content is at least poor
in ferrous iron and represents perhaps a pure Mg-chlorite.

1.2   Chemical and mineralogical composition of
the clay fraction of the soils

In order to establish the clay mineralogical composition
of the saprolites, two main approaches were used. Firstly

X-ray investigations were carried out with a Guinier-de Wolf camera for non-oriented samples, and X-ray diffracto-grams were made of oriented samples. X-ray diffraction peaks can be measured in order to make accurate estimates of the relative amounts of the constituents when dealing with mixtures of a few well crystallized and pure sub-stances (Brindley 1961). Often in soils differences in crystallinity, particle size and chemical composition of the minerals play an important role and severely limit the use of this method. As the weathering materials in the Haarts catchment have different ages, composition and origin, the X-ray diffraction peak measurements on Mg-ethylene glycol clay separates failed to indicate the weathering sequence(s) for the complete profiles. This method could only be used to give semi-quantitative results within the various slope deposits. Due to the complexity of the weathering materials and a mineral assemblage which is difficult to establish semi-quantitatively, the results obtained by X-ray measurements will mainly be treated qualitatively.

Secondly, the elemental composition of the clay fractions were transformed into normative mineralogical composition with the geothite norm (Van der Plas and Van Schuylenborgh 1970). The normative mineralogical com-position can be useful to compare various samples within soil profiles if their mineral assemblage is not too compli-cated. Especially when weathering has proceeded sufficiently far, so that relatively simple mixtures of known weathering products remain, this approach may indicate trends in the transformation of the clay minerals. Here again due to the polygenetic character of most of the soil profiles in the Haarts catchment, a semi-quantitative evaluation of the weathering products for complete profiles is not permitted. The normative mineralogical composition ("norm") does not necessarily agree with the actual mineralogical composition ("mode"). The norm minerals are idealized compounds ("end members") which do not show, like true minerals do, wide ranges in chemical composition. Although the geothite norm calculation can result in rather complex clay mineral assemblage, it should be noted that it does not allow a choice to be made between smectite, chlorite and vermiculite without setting rules about the amount of each compound (Buurman *et al* 1976). These minerals were all formed from MgO, FeO, Al_2O_3 and SiO_2 in varying amounts and in each calculation it is possible to create all of these three minerals, in amounts which are completely dependent upon each other. From the X-ray investigations it could be established that smectite is only present in very minor amounts in the weathering materials and therefore it was ignored in order to calculate the normative chlorite or vermiculite contents.

From X-ray analysis results it was decided, whether chlorite or vermiculite should be included in the norm calculations. This means that chlorite was used in the subsurface soils (buried older slope deposits) and vermi-culite in the surface soils (young slope deposits) and soils on plateau tops. X-ray analysis indicates that

60

chlorite is still present in the surface soil, but it is
hardly possible or even permissible to calculate relative
amounts of the two minerals. The geothite norm itself
does not leave any choice, but ends in an invariable set
of norm minerals (Buurman et al 1976).

1.2.1 Soils on the plateau tops

(profile Haarts 2; see appendix II)

The less than 2 μm fraction of the powdered,
slightly altered parent rock of these soils exists almost
completely of dioctahedral K-mica (10 Å peak) with minor
amounts of chlorite (14 Å, remains upon 550°C heating),
mixed-layer K-mica/chlorite (12 Å spacing, no shift after
Mg-ethylene glycol treatment, remains after 550°C heating
or K-saturation), quartz, hematite, albite and rutile.
The clay fractions of the soil horizons are characterised
by a strong dominance of K-mica. The 10 Å peak is somewhat
smaller in the clay fraction of the surface horizon,
indicating that illite is transformed. The clay fractions
of both soil horizons show clear 14 Å and 12 Å peaks (Mg-
saturated clays). Both peaks diminish strongly upon K-
saturation and 300°C and 550°C heating, and almost no
shift occurs upon solvation (see appendix II, fig 31).
This indicates that vermiculite and a mixed-layer illite-
vermiculite have been formed by weathering, but that also
chlorite and an illite/chlorite interstratification still
exist. The decrease of the 14 Å peak upon K-saturation
is somewhat smaller than upon 550°C heating, indicating
that an Al-hydroxy interlayered mineral is present, which
is also corroborated by a shoulder at the low angle side
of the 10 Å peaks upon 300°C heating (see appendix II,
fig. 31). Al-hydroxy interlayering in these soils is very
limited, probably due to the high organic matter content in
these soils (Rich 1968). Upon solvation with ethylene
glycol the clay fractions of both horizons show a very
small 16 Å peak, which is somewhat more evident in the
lowest horizon (see appendix II, table 40). These 16 Å
spacings may be the result of a clay vermiculite
(Walker 1975) or a vermiculite-smectite interstratification.
The amount of smectite (17-18 Å spacing Mg-ethylene gycol
samples) is very limited in both horizons. Kaolinite was
also established in both clay separates after a treatment
with dimethyl sulfoxide (11.2 Å spacing). Quartz,
geothite, hematite, albite and rutile are present in minor
amounts. These data suggest that some illite is transformed
into vermiculite with an illite/vermiculite interstratification as an intermediate stage. A minor part of the
vermiculite is incompletely filled with Al-hydroxides,
resulting in a limited amount of Al-hydroxy interlayered
mineral. Obviously chemical weathering resulted in a
weathering sequence of dioctahedral micaceous clay minerals
which is often described for weathering under gently acid
conditions in a temperate humid climate (Jackson & Sherman
1953; Jackson 1964; Millot 1964; Sevink 1974).

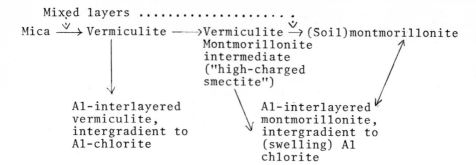

Mixed layers ....

Mica $\xrightarrow{\vee}$ Vermiculite $\longrightarrow$ Vermiculite $\xrightarrow{\vee}$ (Soil)montmorillonite
Montmorillonite
intermediate
("high-charged
smectite")

Al-interlayered   Al-interlayered
vermiculite,   montmorillonite,
intergradient to   intergradient to
Al-chlorite   (swelling) Al
                         chlorite

Scheme of the weathering sequence of micaceous clay
minerals (after Jackson & Sherman 1953, and other
sources: from Veen 1970).

The chemical data and the resultant geothite norm
(appendix II, tables 38 and 39) of these soils also
indicate that the degree of chemical alteration is small.
This is presumably due to the short period of pedogenesis
in these soils. The low ferrous iron/magnesium ratio of the
slightly altered parent rock is maintained in the clay
fractions, indicating that Al-hydroxy interlayering plays
a very subordinate role: relatively important amounts of
ferrous iron are trapped in these interlayer positions
(Carstea *et al* 1970; Brinkman 1977). The vermiculite
formation and/or K depletion of the illite interlayer
positions is suggested by the increase of the CEC of the
clay fractions (after H_2O_2 pretreatment (see appendix II,
table 40), although the CEC remains low due to the high
illite content with almost no or few expandable layers.
Ormsby & Sand (1954) showed a significant linear
relationship between CEC and percent expandable layers in
illites. They concluded that illite with all layers con-
tracted would have a CEC of 15 meq/100 g.

1.2.2   Soils on the steep lower slopes

(profile Haarts 1;   see appendix II)

These soils are polygenetic and have been developed
in three slope deposits. The clay mineralogical composition
of these soils will be discussed according to this
threefold character.

The clay fractions of the grèze-litée like slope
deposit and the pre-Allerød slope deposit with the argillic
horizon are completely identical, with the exception of the
uppermost part of the latter. K-mica (illite) strongly
dominates the clay fractions with very minor amounts of
chlorite, mixed-layer illite/chlorite, quartz and hematite.
Also some kaolinite could be established by the dimethyl
sulfoxide treatment (11.2 Å reflection) and is also
suggested by the intensity ratio of the 14 Å/7 Å reflections
(see appendix II, fig. 33). The identical clay mineralogy
of both older slope deposits, which hardly deviates from
that of the less than 2 µm fraction of the powdered,

slightly altered rocks, indicates that both deposits have
undergone the same weathering history; their clay
mineralogy is even more closely related to the original
lithology than to the present-day plateau soils. Peak
area percentage measurements of the X-ray diffractograms
do not show any transformation of K-mica into vermiculite
and/or smectite (appendix II, table 58). Nevertheless, a
gradual increase of the CEC of the clay fraction going
upward in these saprolites points to a removal of some
potassium from the interlayer positions of the K-mica,
resulting in a more "illitic" character of this mineral.
The upper part of the older slope deposit with the remnant
of the argillic horizon deviates to some extent from the
lower part. A slight decrease in the 10 Å reflection has
been observed, while 16 Å and 17-18 Å reflections appear
(Mg-ethylene glycol treatment), and 14 Å and 12 Å reflections
increase (mg-saturation). Upon K-saturation 12 Å and 14 Å
reflections diminish, indicating that illite/vermiculite
interstratification and vermiculite are present in the
upper part of the argillic horizon. Also a slight decrease
of the 10 Å reflections (Mg-ethylene glycol) could be
established (appendix II, table 58).

The identical clay mineralogical composition of the
two older slope deposits is confirmed by the chemical
composition and the resultant normative mineralogical
composition (appendix II, table 54 en 55; II and III
horizons). All of the normative minerals are present in the
same amounts, indicating that no important alteration has
taken place. The small changes in the clay mineralogical
composition in the upper part of the argillic horizon as
shown by a decrease of the illite content and an increase
in the kaolinite content results from subrecent weathering
"through" the overlying younger slope deposit. The norm
calculations resulted in a slight increase in the free
silica (quartz) and chlorite (= chlorite + vermiculite)
content.

The clay mineralogical composition of the Dystrochrept
in the young slope deposit deviates from that of the older
ones. Although illite still dominates the clay fraction
this mineral is present in smaller amounts. All of the
other minerals detected in the clay fraction of the older
slope deposits are also present. However, 14 Å reflections
(Mg-saturation) appear in considerable intensities which
increase somewhat towards the surface, with the exception
of the surface horizon. These 14 Å peaks almost completely
disappear upon 550°C heating, indicating that vermiculite
and/or Al-hydroxy interlayered mineral strongly dominate
the trioctahedral chlorite. Vermiculite equals or slightly
dominates the Al-hydroxy interlayered mineral in the
lowest part of the Dystrochrept (upon K-saturation
collapsing considerably towards 10 Å; 300°C heating less
important shoulder ranging from 11.5-13.6 Å spacings),
but the reverse is true in the clay fractions of the higher
part, with the exception of the surface horizon of the
Dystrochrept. The dominance of the Al-hydroxy interlayered
mineral over vermiculite is clearly demonstrated in the
plateau-like reflections, in an important shoulder for a

range of 11.7-13.4 Å spacings upon heating to 300°C, or
in a small decrease of the 14 Å reflections upon K-
saturation. In the surface horizon vermiculite is more
important than the Al-hydroxy interlayered mineral,
probably due to hindering of interlayer filling with Al-
hydroxy ions by complexing aluminium with organic
compounds (Rich 1968). In the young slope deposit also
small amounts of smectite (17-18 Å spacing) and
"Clayvermiculite" or vermiculite/smectite interstratification
(16 Å reflection) upon solvation, and kaolinite (11.2 Å
spacing upon dimethyl sulfoxide treatment) could be
established. So it can be concluded that, like in the soils
on the plateau tops, the same mineral weathering sequence
could be observed in the Dystrochrepts. However,
vermiculite and Al-hydroxy interlayered mineral are
present in much higher quantities than in the soils on the
plateau tops. Obviously, chemical weathering seems to be
more effective on the steep lower slopes due to the
different environmental conditions. Also some influence
of a previous weathering cycle, operating during formation
of the alfisols, cannot be completely excluded (young
slope deposit was derived from the truncated horizon of the
alifsols). Nevertheless, it is believed that the clay
mineralogical assemblage of the Dystrochrept is mainly
a result of (sub)recent chemical weathering, which also
influenced the uppermost part of the buried thick slope
deposit.

The chemical composition of the clay fractions and
their derived molar ratios of the soil profile support the
conclusions given by the X-ray investigations. The K_2O/MgO,
FeO/MgO, SiO_2/K_2O, FeO/TiO_2 ratios of the clay fractions
are almost constant in the grèze-litée like slope deposit
and in the intermediate one, indicating their identical
clay mineralogy and lack of quantitatively important
mineral transformation. The upper part of the buried
horizon has a clearly lower K_2O/MgO ratio than the lower
part. However, the former is still higher than those of
the upper slope deposit, indicating some transformation
of illite and its intermediate position between the clay
fraction of the older and younger slope deposits. The
chemical composition of the clay fraction of the Dystrochrept is characterised by lower K_2O/MgO, SiO_2/TiO_2, considerably higher FeO/MgO and FeO/TiO_2 and slightly higher
SiO_2/K_2O ratios. The increase in ferrous iron in this
part of the soil profile has to be attributed mainly to
the formation of an Al-hydroxy interlayered mineral,
because the ferrous iron content of vermiculite is not
significantly higher than that of illite (Weaver & Pollard
1975). Ferrous iron can be built up in these interlayer
positions in relatively considerable amounts. The lower
K_2O/MgO ratios corroborate the X-ray investigations that
in the surface soil K-mica (illite) has been transformed
into vermiculite. The geothite norm calculations
(appendix II, table 55) of the surface soil material confirm

64

this weathering sequence. The normative illite content decreases towards the surface, while normative kaolinite increases. MgO and FeO were completely used for the formation of vermiculite, although from X-ray investigations it could be established that very minor amounts of trioctahedral chlorite occur. The increase in the normative kaolinite content towards the surface is, with the exception of the surface horizon, attributed mainly to the increasing content of the Al-hydroxy interlayered mineral. An increasing kaolinite content cannot be excluded, but this could not be evaluated, due to the small amounts of this mineral and/or its low degree of crystallinity.

In appendix II also the data profile Scheissgrond 1 are presented. Although this soil profile lies in the adjacent Upper Birbaach catchment (fig. 1), it is situated on a similar site as profile Haarts 1. From this profile only the Dystrochrept in the younger slope deposit and the upper part of the buried argillic horizon of the older slope deposits were sampled. X-ray investigations, chemical data, the resultant normative mineralogical composition and micromorphological investigations give similar results as for profile Haarts 1 and therefore this profile will not be discussed in detail. Like profile Nothum 1 (see following section), this profile is developed in two slope deposits derived from parent rock material with a chlorite that is relatively rich in ferrous iron. The chemical data suggest that, if the K_2O/MgO (or $K_2O/(MgO+FeO)$) ratios of the clay fractions and the silt and sand fractions are compared, this chlorite relatively rich in ferrous iron is transformed by chemical weathering to at least the same extent as illite (K_2O/MgO ratios are slightly increasing or constant towards the surface). In contrast the soils developed in the slope deposits or on parent rocks containing Mg-chlorite (profiles Haarts 1, Haarts 2) show, with the exception of the surface horizon of profile Haarts 1, decreasing K_2O/MgO ratios in the clay and non-clay fractions, suggesting that this Mg-chlorite seems to be relatively more stable than illite upon chemical weathering in this environment.

1.2.3 Soils on the gentle upper slopes

(profile Nothum 1; see appendix II)

The clay fraction of these soils shows to some extent a comparable clay mineralogy as that of the soils on the steep lower valley slopes. Here also two slope deposits have been distinguished by micromorphological and palynological investigations. However, from the various depth functions of the soil profile (fig. 13) it could be established that chemical weathering and soil formation which resulted in a Dystrochrept in the young slope deposit also clearly affected the underlying slope deposit. This is shown by the low base saturation and by the Al dominance at the adsorption complex throughout the profile (see appendix II, table 43). Thus it might be expected that also the clay mineral assemblage of the buried older slope deposit with the argillic horizon has been altered by the (sub)

recent weathering process, and due to this complexity
these soils are treated last.

X-ray diffractograms (appendix II, table 49) and
Guinier-de Wolf films indicate that illite dominates the
clay fraction, even if all of the intergrades are taken
into account. A decrease in the illite content towards
the surface could be established, indicated by smaller
10 Å peaks and lower K_2O contents (lower normative illite

content) (appendix II, table 46). This decrease in the
illite content is accompanied by an increase in the 14 Å
reflections, which almost completely collapse to 10 A upon
K-saturation and heating to 300^0C and 550^0C. This means
that vermiculite is the dominant 14 A mineral, although
also an Al-hydroxy interlayered mineral and trioctahedral
chlorite are present throughout the profile (appendix II,
fig. 32). In the clay fraction of IIB2tb, the lowest
horizon of this profile, hardly any Al-hydroxy interlayered
mineral was observed and also its vermiculite content is
less than in the older clay fractions of this soil profile.
Trioctahedral chlorite is present throughout the soil profile.
Its contribution to the 14 A reflections (Mg-saturation)
diminishes gradually towards the surface. Kaolinite could
be detected throughout the profile upon dimethyl sulfoxide
saturation (11.2 Å spacing). Also the 14 A/7 A ratio, the
presence of a relatively important 7 A peak upon heating
to 300^0C and the complete disappearance of this peak upon
550^0C heating point to the presence of this mineral. No
trend regarding its amounts throughout the profile could
be established.

The composition of the clay fractions of profile
Nothum 1 slightly deviates from those of profile Haarts 1
and Scheissgrond 1 only by the relatively more important
16 Å and 17-18 Å reflections upon solvation, indicating a
somewhat higher clay vermiculite or vermiculite/mont-
morillonite and, to a lesser degree, montmorillonite content.
So it can be concluded that, with minor deviations in this
soil profile, a similar weathering sequence occurs as in
the other soils of the Haarts catchments. Due to the fact
that the buried slope deposit is covered by a relatively
thin surface mantle, its clay mineralogical assemblage
is more influenced by the (sub)recent weathering processes,
resulting in a higher vemiculite and a lower illite content
with respect to the clay fractions of the buried thick
older slope deposit of the profile Haarts 1.

The chemical composition of the clay fraction and the
geothite norm (appendix II, table 45 and 46) indicate that
some illite is transformed. However, the K_2O/MgO and the

$K_2O/(MgO+FeO)$ ratios of the clay fractions, and also those
of the non-clay fractions, slightly increase towards the
surface, suggesting that chlorite relatively rich in ferrous
iron is attacked more rapidly than illite by weathering,
like in profile Scheissgrond 1. Here also the chlorite in
the clay fraction has a relatively high FeO/MgO ratio, which
is inherited from the parent rock. The FeO/MgO ratios of
the clay fraction throughout the profile show relatively

minor changes with respect to the other soil profiles
(0.30 → 0.20 → 0.27), corroborating the X-ray analysis
investigations that Al-hydroxy interlayering plays only a
subordinate role in this soil profile. The geothite norm
(appendix II, table 45) gives constant normative kaolinite
contents, except in the lowest horizon, suggesting the
minor importance of this Al-hydroxy interlayering and the
lack of any trend regarding its amounts throughout the
soil profile.

1.3   Chemical and mineralogical composition of the
      sand and silt fractions of the soils

     The chemical and mineralogical investigations of the
sand and silt fractions of the various soils give, with
some exceptions, similar results. Therefore no separate
treatment of these soil profiles will be given. X-ray
investigations established that the mineralogy of the non-
clay fractions is mainly identical to those of the slightly
altered rocks. Quartz, K.mica(illite), albite, chlorite
and a mixed-layer illite/chlorite could be detected with
minor amounts of hematite, geothite and rutile. Also
vermiculite and an Al-hydroxy interlayered mineral appear
in the non-clay fraction of those soil horizons, where
these two minerals are also present in the clay fractions
(surface soils). The chemical composition and the epinorm
(Burri 1964) give some indications of the weathering of
the non-clay fractions. Normative free silica (quartz) is
residually enriched and normative albite slightly decreases
towards the surface, suggesting alteration of albite. The
normative muscovite content in the epinorm of the various
soil profiles decreases towards the surface, indicating
that K-mica (illite) is transformed. This is especially
the case in the non-clay fractions of the soils of the
younger slope deposits and the young plateau soils, and in
the uppermost part of the buried slope deposit. In the
soil profiles where chlorite is present as a Mg-chlorite
poor in ferrous iron, the K_2O/MgO ratios of the sand and
silt fractions also suggest that this chlorite has been
attacked to a much lower extent by chemical weathering
than illite (profile Haarts 1 and 2). If the non-clay
fractions contain a chlorite relatively rich in ferrous
iron (profile Nothum 1, Scheissgrond 1), increasing K_2O/MgO
ratios towards the surface have been established, like in
the clay fractions, probably caused by a more rapid transformation of this chlorite than of illite. The higher
FeO/MgO ratios in the non-clay fractions of the surface
soils of the slope deposits and of the young plateau soils
are mainly attributed to the Al-hydroxy interlayering with
ferrous iron "trapped" in the interlayer positions in the
silt and sand fractions (Carstea et al 1970a; Coffman &
Fanning 1975; Brinkmann 1977) and to the influence of
organic matter. The higher Fe-chlorite content in the
epinorm calculations for the surface horizons of the surface
soils (appendix II) do not indicate that primary chlorite
is residually enriched in this part of the soils. It is
caused because in epinorm calculations all of the FeO is
attributed to chlorite.

2. CHEMICAL COMPOSITION OF THE LIQUID PHASE

2.1 Chemistry of atmospheric precipitation

2.1.1 Introduction

2.1.1.1 General

The chemical and physical characteristics of precipitation coming into contact with rock or soils have often been ignored by soil scientists and geomorphologists. A possible reason for this is the relatively high dissolved matter content commonly found in waters of many drainage basins. This high content often effectively masks the chemical contribution of the precipitation component to the geochemical balance. A detailed knowledge of precipitation chemistry, however, is necessary not only in order to evaluate its influence on the rock(soil)-water interactions and to establish the geochemical balance, but also for an accurate estimation of the chemical denudation rates in detailed input-output studies.

The geochemical and pedological significance of cyclic salts was throughouly reviewed by Junge (1958), Carroll (1962), Gorham (1958, 1961), Oden (1976) and in a series of important papers by Eriksson (1955, 1959, 1960). Corrections for these atmospheric contributions are becoming increasingly common in geochemical studies, as for example the Hubbard Brook Experimental Forest, New Hampshire, for nutrient cycling (Bormann & Likens 1967; Fisher *et al* 1968; Likens *et al* 1977) and in other recent work in the United States (Cleaves *et al* 1970; Morehead 1971).

The study of particular ions in precipitation in relation to nutrient cycling is also assuming new importance. The work of Anderson (1945) in Australia, with regard to chlorides has been extended by Schoeller (1961 a, b; 1963 a, b) to calculate the water balance of catchments, as exemplified in the study of annual budgets by Juang & Johnsson (1967) at Hubbard Brook, and Peck *et al* (1973) in Western Australia.

Finally, the increasing acidity of precipitation first observed in Scandinavia (Gorham 1955) and later in the Northeastern United States (Gambell & Fisher 1966; Cogbill & Likens 1974; Likens & Bormann 1974; Galloway *et al* 1976a, 1976b; Likens *et al* 1976; Hornbeck *et al* 1977) and now shown to be directly attributable to industrial activity (Bolin 1971; Granat 1972 a and b; Gorham 1976; Vermeulen 1977; Ridder 1978) has resulted in a much better understanding of the chemistry of precipitation and its influence upon ecosystems (Granat 1972a; Cogbill & Likens 1974; Brosset 1976; Oden 1976).

In this section input of nutrients in the area investigated in relation to several parameters affecting it will be given and the chemistry of the precipitation will be discussed.

2.1.1.2   Components of precipitation

The component of precipitation considered here is the "wet component" of rain and snow which is collected in the rain gauges during rain or snow events. However, not only this component contributes to the deposition of nutrients, because there is also a dry component that is transferred from the atmosphere to the ground surfaces. This component is termed dry deposition and has three subcomponents (Galloway & Likens 1978):

Dry fallout: particles that are affected by gravity to such a degree that they fall on earth surfaces (vegetation, water, soils and rocks). These particles are necessarily large (mostly > 2 µm diameter) and are usually derived from soils, plant debris or condensed aerosols.

Impacted aerosols: smaller particles that are impacted onto earth surfaces.

Adsorbed gases: gases that are adsorbed on or by earth surfaces.

Due to the difficulties in sampling atmospheric aerosols and dry fallout, no results were obtained on their contribution. Only the wet deposition, for various precipitation events and the bulk precipitation (wet and dry deposition) for the "monthly" samples could be established. It has also to be taken into account that the bulk precipitation samples were analysed after filtration. Consequently only the water soluble part of the dry deposition and wet deposition were traced in these samples.

2.1.1.3   Problems of estimating the chemical composition of (bulk) precipitation samples

This section briefly discussed problems which have to be solved or taken into account in order to get a (relatively) uncontaminated and representative precipitation sample. For a more comprehensive review the reader is referred to Galloway & Likens (1976, 1978), Vermeulen (1977) and Ridder (1978).

The ideal precipitation collector should representatively sample precipitation according to precipitation composition (concentrations) and ecosystem loading (input). However, many complications have to be faced, notably: site and altitude of the precipitation collectors, collector and collecting vessel material, efficiency of the collectors, contamination by bird faeces, leaves, bud scales, insects, soot, algae, the adsorption of various elements from the collector and vessel walls, the influence of dry fallout, sampling period and sample preservation and -storage.

In general it can be assumed that a collection site is not influenced by local potential sources of atmospheric debris such as roads, marshes, smoke stacks etc. If there are local sources, the wind directions have to be investigated during precipitation events and sampling periods. The site of the collectors has to be on a flat or gently sloping terrain and sheltered from the wind. A clearing

69

in a forest with a grass cover is ideal. The collector
has to be protected against splashed material from the
ground surface. This can be done by installing the
collector at more than 100 cm above the ground surface. To
establish the representativeness of the collector a number
of identical collectors should be placed on a grid pattern
around the permanent collector for a certain period. For
the determination of the inorganic composition of
precipitation, the collection system should be of (high
pressure) polythene or teflon in order to suppress ion
exchange reactions. Adsorption of elements against the
collector walls might be important if heavy metals are
investigated, but this phenomenon can be ignored if the
sample is acidified with nitric acid to pH~1.5. The
latter is not relevant for this study because no heavy
metals were quantitatively detected in the samples. If
the investigation is designed the atmospheric inputs into
the ecosystems the bulk deposition may be adequate. If
the data are to be used to investigate the chemistry of
rain or snow, then the dry deposition has to be excluded
from the samples. There are also several factors that
determine the length of sampling period. Stability of the
sampling is the most important factor (Galloway & Likens
1978). If a reactive parameter in the precipitation is to
be determined, like organic acid, then event sampling is
necessary. However, if a conservative property is to be
measured, such as precipitation volume (if the diameter
of the inlet to the polythene bag is small) or if the
sample is resistant to chemical change (most rural
precipitation samples with pH < 4.8 are essentially self-
preserving (Galloway & Likens 1978)), then the sampling
schedule may be on a weekly or even on a monthly basis if
black painted polythene bottles are used. However, there
is a strictly limited period for which samples can be left
in the field, especially in the summer months, because
sample contamination or alteration is greatly increased
and no ideal biocide is known at the present time. Finally
the samples have to be analysed immediately or stored in a
dark place at 4^{0}C.

Due to logistic problems the above requirements could
not completely be fulfilled. Samples had to be collected
monthly. Only when the author was in the field data on
precipitation events could be obtained or samples on a
weekly basis be collected.

Data on precipitation chemistry are also available for
a site relatively nearby in Belgium. These data were
collected until 1974 and analysed for the years 1967-74.
The data are of very poor quality (often unbalanced and
unreliable data) due to contamination, analytical- and
presentation errors (see also Ridder 1978). For this
reason these date are of very limited value and can only be
used for a very general comparison with the samples from the
area investigated.

2.1.2 Results

2.1.2.1 General

The chemical composition of the bulk
precipitation, ie. a mixture of rain, snow and dry fallout
(Whitehead & Feth 1964) for the catchment is given in the
appendix III, table 77. The weighted concentrations and
total input of chemicals in bulk precipitation on a sample
period basis are given in table 9 and 10. Various samples
were obviously contaminated with bird feces, insects,
leaves, pollen, soot, etc. This contamination represents
a serious problem especially for the elements hydrogen
(ammonium), nitrogen, potassium and the alkalinity and
acidity of precipitation samples. Therefore two weighted
average concentrations and total input values are given in
table 9 and 10. One for all samples (clean and dirty) and
one for only clean samples. The input on a clean sample
basis for the whole period was calculated by multiplication
of the amount of precipitation for this period (160 .5 mm)
and the weighted average concentration of the clean
samples. It had to be assumed that the weighted average
concentration of the clean samples is representative for
the whole period. To calculate the input on the basis of
only the clean samples seems realistic, in view of the very
good balance of the cations and anions. Also the weighted
average concentrations of some non-coastal 1 cations from
Belgium give similar results. The input data show that
significantly large amounts of nutrients may be added to
ecosystems in bulk precipitation. This is particularly
true for sulphur, nitrogen and chloride. In table 9 some
additional data are given on the concentration levels of
the various ions in the precipitation of the catchment.

2.1.2.2 Wind direction, precipitation
and ion supply

From fig. 17 and table 12 can be seen that during
the investigation period there was a bimodal distribution
in the frequency of wind directions with the NE and
westerly components clearly dominant. However, the distribution of rainday winds is quite different. For those days
the westerly components are definitely dominant over all
other components. This dominance of westerly winds is more
evident when the amount of precipitation is considered,
because most rainfall is ordinarily associated with the
passage of fronts, indicating that the westerly component
is greatest at these times.

The Haarts catchment is situated in the central part
of NW Europe, and industrial areas lie to the N, NE, E, SE
and S in Belgium, Western Germany, Southern Luxembourg and
France. Thus for the catchment it is possible that
industrial pollution may be a very important source of
cyclic ions. This is clearly illustrated by the significan
negative correlation with α chosen as 0.05 (α = level of
significance (Siegel 1956)) between the frequency of the
westerly winds, the SW NW quadrant in which the industrial

Table 9   Chemical concentrations in mg/l and frequency of wind direction during the collection periods in the Haarts catchment

Period	Precipitation	pH	average concentrations mg/l											Frequency of wind directions (percentages)							
			H	K	Na	Ca	Mg	NH$_4$	NO$_3$	Cl	SO$_4$	HCO$_3$	N	NE	E	SE	S	SW	W	NW
9/10-5/11 '73	84.7	4.50	0.03	0.16	0.51	1.04	0.02	0.76	1.55	2.09	4.13	-	10.8	15.7	13.3	16.9	4.8	3.6	20.5	14.5
5/11-10/12	70.2	5.53	tr	0.08	0.28	0.21	0.01	1.47	0.12	3.24	1.62	-	13.3	11.4	4.8	1.9	1.0	16.2	16.2	35.2
10/12-7/1 '74	66.5	4.48	0.03	0.20	1.13	1.24	0.01	1.07	0.96	3.32	1.92	-	4.8	8.3	15.5	19.0	9.5	20.2	13.1	9.5
7/1-11/2	95.8	4.22	0.06	0.39	0.62	0.52	0.01	0.79	1.33	2.77	3.65	-	4.8	3.8	-	15.2	17.1	32.4	16.2	10.5
11/2-18/3	40.3	4.00	0.10	0.20	0.37	1.16	tr	1.64	3.22	1.49	4.51	-	4.8	35.2	11.4	13.3	6.7	5.7	15.2	7.6
18/3-22/4 (*)	26.5	4.27	0.05	0.78	1.01	2.90	0.30	0.80	11.78	3.27	10.95	-	3.8	63.8	21.0	4.8	1.9	1.9	0.9	1.9
22/4-30/5	58.4	5.33	tr	0.90	0.62	1.34	0.15	1.62	1.80	2.45	8.65	-	16.7	20.2	17.5	8.8	4.4	6.1	7.9	18.4
30/5-1/7	74.3	4.92	0.01	0.55	0.53	0.80	0.17	0.83	1.32	1.56	3.99	-	14.6	16.7	9.4	7.3	8.3	15.6	15.6	12.5
1/7-8/8*	71.4	6.44	tr	2.14	0.42	0.36	0.19	6.07	0.52	1.99	5.81	18.31	14.0	5.3	0.9	4.4	9.6	16.7	21.9	27.2
8/8-22/10 (*)	252.8	5.79	tr	0.58	0.50	0.48	0.17	1.93	0.85	1.78	7.56	7.58	7.0	13.6	9.2	9.6	7.9	24.6	12.7	15.4
22/10-19/11	73.4	4.96	0.01	0.15	0.76	0.58	0.06	0.45	1.45	0.96	2.27	-	12.3	3.7	3.7	2.5	13.6	23.5	16.1	24.7
19/11-18/12	107.9	4.47	0.03	0.08	0.82	0.41	0.12	0.47	1.45	2.11	2.44	-	7.4	1.2	-	7.4	6.2	25.9	28.4	23.5
18/12-21/1 '75 (*)	69.5	4.70	0.02	0.39	0.35	0.94	0.17	0.31	1.09	1.60	4.42	-	4.9	-	-	7.8	14.7	31.4	24.5	16.7
21/1-18/2	64.4	4.74	0.02	0.14	1.01	0.50	0.20	0.25	0.57	1.33	1.80	-	10.7	11.9	11.9	9.5	10.7	19.0	14.3	11.9
18/2-20/3	69.8	4.60	0.03	0.23	0.66	1.13	0.13	0.72	3.03	0.90	5.25	-	15.6	20.0	23.3	8.9	14.4	6.7	7.8	3.3
20/3-12/5	99.0	4.00	0.10	0.12	0.37	0.54	0.01	0.98	2.91	0.89	2.59	-	20.8	18.2	9.4	6.3	1.9	14.5	15.7	13.2
12/5-16/7 (*)	103.8	6.10	tr	0.76	0.38	0.45	0.10	2.90	1.16	0.50	5.10	6.00	19.6	35.2	12.6	7.5	3.5	8.0	5.5	8.0
16/7-26/8*	73.1	6.39	tr	0.78	0.37	0.46	0.12	2.94	1.18	0.35	6.92	7.02	11.4	17.1	13.8	6.5	5.7	11.4	8.1	26.0
26/8-15/10 (*)	104.7	5.80	tr	0.77	1.00	1.25	0.16	1.96	5.64	1.60	4.51	8.60	10.6	23.8	10.6	9.9	7.3	21.9	7.3	8.6
9/10'73-15/10'75	1606.5	4.66	0.02	0.49	0.60	0.73	0.12	1.54	1.81	1.73	4.39	3.27	11.5	18.1	10.1	8.6	7.5	15.6	13.7	15.0
clean samples °	904.7	4.43	0.04	0.25	0.67	0.74	0.07	0.97	1.62	1.93	3.94	-	12.1	14.2	10.1	9.4	7.9	14.6	16.0	15.6

(*) slightly dirty samples, * dirty samples, ° weighted mean

72

Table 10   Inputs by sample periods in g/ha in the
            Haarts catchment

Period	H	K	Na	Ca	Mg	NH$_4$	NO$_3$	Cl	SO$_4$	HCO$_3$
/10-5/11 '73	27	136	432	883	17	644	1313	1772	3498	-
/11-10/12	2.1	56	197	147	7.0	1031	87	2274	1137	-
)/12-7/1 '74	22	133	751	825	6.7	711	638	2208	1277	-
/1-11/2	58	374	594	498	9.6	757	1274	2654	3497	-
1/2-18/3	42	81	149	467	2.0	661	1298	600	1818	-
3/3-22/4 (*)	14	207	268	769	80	212	3122	867	2902	-
2/4-30/5	2.8	526	362	783	88	946	1051	1431	5052	-
)/5-1/7	9.0	409	394	594	126	617	981	1159	2962	-
/7-8/8 *	0.3	1527	300	257	136	4334	371	1421	4148	13073
/8-22/10 (*)	4.1	1466	1264	1213	430	4879	2149	4500	19112	19162
2/10-19/11	8.1	110	558	426	44	330	1064	705	1666	-
9/11-18/12	37	86	885	442	129	507	1565	2277	2633	-
3/12-21/1 '75 (*)	14	272	240	654	118	213	754	1109	3071	-
1/1-18/2	12	90	650	322	129	161	367	857	1159	-
3/2-20/3	18	160	462	787	89	505	2114	626	3662	-
)/3-12/5	100	119	360	535	10	970	2881	881	2564	-
2/5-16/7 (*)	0.8	789	394	467	104	3010	1204	519	5294	6228
5/7-26/8 *	0.3	570	270	336	88	2149	863	256	5059	5132
5/8-15/10 (*)	1.7	806	1047	1309	168	2052	5905	1675	4722	9004
/10 '73-15/10 '75	373	7917	9577	11714	1863	24689	29001	27791	70511	52599
lean samples	600	4049	10785	11913	1167	13922	25984	30976	63299	-

(*) slightly dirty samples; * dirty samples

activity is at a minimum, and the mean weighted sample
period of sulphate (r = -0.602), nitrate (r = -0.617) and
calcium (r = -0.802) concentrations. For all other ionic
concentrations no significant relationships could be
detected, which can be explained by the great influence on
the chemical composition of the dry fallout and the rather
high frequency of all wind directions (table 12). The
great importance of dry fallout, coming from air masses
supplied by winds from all directions (table 12), in the
bulk precipitation composition is also confirmed by the lack
of relationship between the amount of precipitation and the
element input in the catchment (table 13). Only the sodium
input is significantly correlated with α chosen at 0.10
(r = 0.515) with the amount of precipitation and with the
frequency of westerly winds (r = 0.69) (table 15). This
relationship is easy to understand because almost all
sodium has seawater as its source and the amount of
precipitation is closely related to westerly winds.

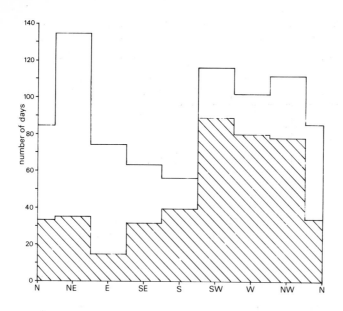

Figure 17   Wind directions on all days (blank and shaded)
and precipitation days (shaded) at Berlé
(9.10.73 - 15.10.75)

Cation ratios may be used to indicate the source of
chemicals in precipitation (Junge 1963). Precipitation
samples from air masses of continental origin usually
show appreciably higher calcium ratios than those of western
origin. This is confirmed by the significant relationships
with α chosen as 0.005 between the calcium/sodium ratios and
the westerly wind components (r = -0.750).

2.1.3   Chemical Data

2.1.3.1   Calcium, magnesium and potassium

Calcium. The concentration of calcium in bulk
precipitation ranged from a maximum of 1.34 mg/1 to a
minimum of 0.21 mg/1, with a mean value of 0.79 mg/1 and
without a pronounced seasonal trend. The enrichment of
calcium in the bulk precipitation compared to seawater
indicated by the higher calcium/sodium ratios, 0.50-3.14
instead of 0.04, is striking. The large decline in sodium
concentrations in precipitation with distance from the sea
coupled with an increase in calcium from soil particles and
industrial pollution produces these large calcium/sodium
ratios. Such ratios were found in both clean and dirty
bulk precipitation samples (table 14). The influence of
the non-coastal sources of calcium is also confirmed by
the significant negative relationship between the frequency
of westerly winds and the calcium concentrations (r =
-0.626) (table 15).

Table 11   Statistics of bulk precipitation chemistry
(concentrations in mg/1)

Ion	Mean value	Minimum value	Maximum value	Standard deviation	$S_{\bar{x}}/\bar{x}$, %
K	0.27	0.12	0.90	0.24	88.9
Na	0.67	0.37	1.13	0.24	35.8
Ca	0.79	0.21	1.34	0.38	48.1
Mg	0.07	0.00	0.20	0.07	100.0
NH_4	0.92	0.25	1.64	0.46	50.0
NO_3	1.64	0.12	3.22	0.97	59.1
Cl	1.93	0.89	3.32	0.88	45.6
SO_4	3.57	1.62	8.65	1.99	55.7
H	0.04	0.00	0.10	0.03	75.0

Table 12   Frequency of wind directions in percentages
at Berle (9.10.73 - 15.10.75)

	N	NE	E	SE	S	SW	W	NW
All days (736)	11.5	18.1	10.1	8.6	7.5	15.6	13.7	15.0
Precipitation days (399)	8.4	8.7	3.8	8.0	9.8	22.1	19.8	19.3

The problem of correlating bulk precipitation calcium
factors is compounded by evidence that a large part of the
calcium is associated with the dry fallout component. For
example Whitehead & Feth (1964) showed that there was a
significant difference between "rainwater" and "bulk
precipitation", particularly for calcium. Gorham (1958)
could establish from analyses of smoke solids that calcium
and chloride form the major part of the smoke. Also
Meszaros (1966) showed that possibly over half of the
calcium in the atmosphere may be in the non-dissolved form
as atmospheric dust, which considerably contributed to the
composition of bulk precipitation.

In the precipitation samples from the Haarts catchment
calcium was highly positively correlated with sulphate,

Table 13 Linear correlation coefficients

x	r_{xy}	r_{xz}
H	0.51*	0.01
K	-0.04	-0.23
Na	0.53*	0.60**
Ca	-0.04	-0.63**
Mg	0.09	0.04
NH_4	0.14	-0.03
NO_3	0.06	-0.29
SO_4	0.17	-0.37
Cl	0.47	0.57*

* $\alpha \leq 0.10$; ** $\alpha \leq 0.05$.

nitrate and potassium. A negative relationship exists with the frequency of non-westerly winds (table 15), although the significance of the correlations is not always clear. These data suggest that calcium is directly related to acids, possibly sulphuric and nitric acids (Granat 1972a; Cogbill & Likens 1974).

Magnesium. The concentrations of magnesium in the bulk precipitation ranged from 0 to 0.20 mg/l, with an average of 0.07 mg/l. So the observed levels were very low compared to the other elements. Magnesium showed poor correlations with all variables (table 15), only a relationship could be found with the temperature, indicating that the magnesium levels were generally higher in the summer periods. The amount of magnesium found in precipitation is on the average equal to that expected from the Mg/Na ratio in seawater. Nevertheless, the Mg/Na ratios are significantly correlated with the temperature (r = 0.687; α chosen as 0.02) indicating a seasonal trend in the magnesium concentrations because no mechanism which selectively washes out magnesium from the atmosphere has yet been demonstrated.

Potassium. Like magnesium, potassium was very variable in concentration throughout the period of investigation. Its concentration varied between 0.12 - 0.90 mg/l with an average of 0.27 mg/l. Potassium concentrations are positively related with the temperature, possibly indicating a seasonal effect, and with calcium and sulphate, although the significance of these last two correlations is not clear (table 15). Whereas in seawater the ratio K/Na is

Table 14   Concentrations (mg/l) ratios for precipitation
            and frequency of westerly wind directions
            (perc.) for collection periods

Period	Ca/Na	Mg/Na	K/Na	Cl/Na	SO_4/Na	SW-NW winds
9/10-5/11'73	2.04	0.04	0.31	4.10	8.10	38.6
5/11-10/12	0.75	0.04	0.29	11.57?	5.79	67.6
10/12-7/1 '74	1.10	0.01	0.18	2.94	1.70	42.8
7/1-11/2	0.84	0.02	0.63	4.47	5.89	59.1
11/2-18/3	3.14	<0.01	0.54	4.03	12.19	28.5
18/3-22/4 (*)	2.87	0.30	0.77	3.24	10.84	4.7
22/4-30/5	2.16	0.24	1.45	3.95	13.95	32.4
30/5-1/7	1.51	0.32	1.03	2.94	7.53	43.7
1/7-8/8*	0.86	0.45	5.09	4.74	13.83	65.8
8/8-22/10 (*)	0.96	0.37	1.16	3.56	15.12	52.7
22/10-19/11	0.76	0.08	0.20	1.26	2.99	64.3
19/11-18/12	0.50	0.15	0.10	2.57	2.98	77.8
18/12-21/1'75 (*)	2.69	0.49	1.11	4.57	12.63	72.6
21/1-18/2	0.50	0.20	0.14	1.32	1.78	45.2
18/2-20/3	1.71	0.20	0.35	1.36	7.95	17.8
20/3-12/5	1.46	0.03	0.32	2.41	7.00	43.4
12/5-16/7 (*)	1.18	0.26	2.00	1.32	13.42	21.5
16/7-26/8*	1.24	0.32	2.11	0.95	18.70	45.5
26/8-15/10 (*)	1.25	0.16	0.77	1.60	4.51	37.8
9/10'73-15/10 '75°	1.22	0.20	0.82	2.88	7.32	44.3
clean samples°	1.16	0.11	0.39	3.02	6.16	46.2
sea water	0.04	0.12	0.04	1.80	0.25	

(*) slightly dirty samples; * dirty samples; ° weighted mean.

Table 15   Matrix of linear correlation coefficients* of precipitation characteristics (n = 12)

	pH	K	Na	Ca	Mg	NH$_4$	NO$_3$	Cl	SO$_4$	Precip.	W-wind	Temp.
pH	1.00	0.35	0.32	-0.13	0.40	0.15	-0.61	0.31	0.15	-0.25	0.21	0.22
K		1.00	-0.18	0.51	0.40	0.40	0.06	0.15	0.83	-0.27	-0.35	0.66
Na			1.00	-0.08	0.30	-0.45	-0.72	0.40	-0.43	-0.02	0.31	-0.44
Ca				1.00	0.04	0.41	0.47	0.05	0.69	0.44	-0.80	0.17
Mg					1.00	-0.40	-0.14	-0.34	0.28	-0.09	-0.13	0.48
NH$_4$						1.00	0.51	0.41	0.47	-0.52	-0.32	-0.01
NO$_3$							1.00	-0.61	0.46	-0.11	-0.62	0.08
Cl								1.00	-0.06	0.01	0.33	-0.25
SO$_4$									1.00	-0.33	-0.60	0.49
Precip.										1.00	0.57	0.11
Wind											1.00	-0.16
Temp.												1.00

* $|r| \geq 0.497$ $\alpha = 0.10$; $|r| \geq 0.576$ $\alpha = 0.05$; $|r| \geq 0.658$ $\alpha = 0.02$; $|r| \geq 0.708$ $\alpha = 0.01$.

0.04 in the bulk precipitation in the Haarts catchment this ratio varies between 0.10-5.09, with an average of 0.46. The only reasonable explanation for this fact is an additional source of potassium from the soil because chemical decomposition of sea-spray particles or, as is sometimes assumed, a mechanical separation of their constituents is very unlikely to increase this ratio by a factor of about 10-20 (Junge & Werby 1958). However, it has to be borne in mind that potassium concentrations of bulk precipitation can be strongly influenced by slight contamination (table 9), although the observed potassium levels of the clean samples are quite reasonable compared with data from other non-coastal areas (Junge & Werby 1958; Gorham 1961; Feth et al 1964; Likens et al 1967; Sugawara 1967).

2.1.3.2 Sodium and chloride

Cations and anions in precipitation originate from several sources, including sea spray, terrestrial dust and gaseous pollutants. The solubility of atmospheric gases from either natural and/or anthropogenic origin in atmospheric water is clear, but the origin of non-volatile components such as Na, Mg, Ca and K in precipitation is not so obvious.

The behaviour of sodium is of considerable interest, because it is commonly assumed that all sodium in precipitation is directly derived from seawater (Gorham 1958; Junge 1959; Gambell & Fisher 1966; Granat 1972a). Therefore the ratios of ions in precipitation to sodium might be indicative of the influence of marine factors on precipitation chemistry (Junge & Werby 1958; Eriksson 1959, 1960). The ratios of calcium, magnesium and potassium to sodium have already been discussed in the previous section, so in this paragraph only the chloride/sodium ratio will be considered. This ratio varies between 1.26 and 11.57, with an average of 3.58 ($\pm$ 2.76), compared to the seawater ratio of 1.80 (table 14). However, if the outlier in this ratio of 11.57, presumably due to an analytical error resulting in a too low sodium content is neglected, the choride/sodium ratio varies between 1.26 and 4.74, with a mean of 2.85 ($\pm$ 1.19). The variability in this ratio is also indicated by a rather poor correlation of sodium and chloride concentrations ($r = 0.403$). So it can be concluded that various factors must have influenced the chloride and sodium levels.

Firstly (I) the chemical composition of sea spray can somewhat deviate from that of seawater. Fluctuations in the chemical composition of the uppermost seawater layer have been established so that the chloride/sodium ratios only 3 metres above the breaking sea surface may be significantly different from those that would be expected (Chesselet et al 1972). Secondly (2) one of the two constituents may be withdrawn faster from a water droplet in the atmosphere. A possible mechanism is the transformation of NaCl into HCl through the action of SO_2 resulting in an escape of HCl from the droplet and a possible increase of sodium in the precipitation (Ewan & Philips 1975). Thirdly (3) other sodium and/or chloride sources are present, which will be relatively more influential when the investigated area is situated on a larger distance from the coast. Junge & Werby (1958) showed for the USA that there is a marked decrease in the chloride and sodium levels proceeding inland due to the dilution of maritime air masses. They also found a considerable decrease in the chloride/sodium ratio from 1.80 to 0.30 going inland, which might be explained by a loss of chloride caused by decomposition of sea-spray particles or by the supply of additional sodium (terrestrial dust). Sodium and potassium occur in soils in temperate humid climates primarily as silicates and will be converted into more soluble salts only within the soil and predominantly in moist areas. Since there can be little doubt, however, that most of the "excess of potassium" must come from the soil, it seems reasonable to explain at least a part of the "excess sodium" in precipitation by the same source. It would seem that if processes (2) and especially (3) are effective, they will result, if industrial chloride input is low, in relatively low chloride/sodium ratios.

Nevertheless, for the greater part of the year the chloride/sodium ratio is much greater than 1.80, suggesting that at least part of the chloride is derived from terrestrial sources. Especially in urban areas appreciable amounts of chloride, mainly due to combustion of coal, have been detected (Gorham 1958). Therefore an "excess chloride" in the precipitation in the Haarts catchment, which is situated not too far away from industrial areas, is not surprising, although the sources are conjectural.

The chloride concentrations ranged from 0.89 to 3.32 mg/l with an average of 1.93 mg/l (table 11). The concentrations do not show any significant relationship with other variables (table 15) with the exception of nitrate. The twofold origin of the chloride, eg. industrial pollution and sea spray is also corroborated, not only by the high "excess chloride" levels, but also by a significant correlation with α chosen an 0.10 of the chloride input with the frequency of westerly winds (r = 0.572).

Sodium concentrations ranged between 0.37 and 1.13 mg/l with a mean of 0.67 (± 0.24) mg/l. They are significantly negatively correlated with nitrate (r = -0.717), an element of completely terrestrial origin. The sodium input in the catchment shows a close relationship with the frequency of westerly winds (r = 0.686), indicating that most of the sodium has a maritime origin.

2.1.3.3 Nitrogen and sulphur

The results for ammonium, nitrate and sulphate are of particular interest in view of their quantitative importance (see table 9), with the exception of some cases in urban areas where chloride is important. The excess concentrations of these components almost completely govern

the acidity and alkalinity of the precipitation. For this
reason these constituents,which in Western Europe are
mainly derived from terrestrial (industrial) sources, are
of very great importance for the ecosystems.

Nitrogen. Nitrogen is the most important element in
the atmosphere; it includes N_2, N_2O, NO_x, NH gas, and
ammonium (NH_4^+), nitrate (NO_3^-) and organic nitrogen
aerosols. Of the eight possible oxides of nitrogen, only
N_2O, NO and NO_2 are important constituents in the
atmosphere, of which the relatively inert N_2O is most
abundant. Only NO_2, NO_3^-, NH_3, NH_4^+ and part of the organic
nitrogen is readily soluble in water (Söderlund & Svensson
1976). A large proportion of the NH_3 and NO_x in the
atmosphere may originate from biological processes, but
also industrial activities, traffic and urine of animals
for instance at dairy farms, will affect the concentration
levels. Nitrogen dioxide can react with water vapour to
form nitric acid, which, in turn, reacts with NH_3 or
particles in the air to form nitrate salts, like NH_4NO_3,
which can rain out or wash out from the atmosphere.
The major reactions involving the various gaseous nitrogen
compounds are the oxidation reactions of NH_3 and NO_2 to
form nitrates and the neutralization of NH_3 to form
$(NH_4)_2SO_4$,NH_4HSO_4 and NH_4NO_3 aerosols, which are removed
from the atmosphere by precipitation (Tabatebai & Laflen
1976). The residence times for all nitrogen compounds
except N_2 and N_2O are only a few days. For N_2O the
residence time has been estimated as being less than 200
years, if there is no loss in the biosphere. Biological
reactions, however, may reduce this residence time to about
1 to 3 years (Robinson & Robbins 1970).

In (bulk) precipitation, the quantitatively important
nitrogen compounds are ammonium and nitrate. The concentration levels of these compounds indicate that both
ions are important constituents in the precipitation in the
Haarts catchment. The concentration levels for ammonium
ranged from 0.25 to 1.64 mg/l, with a mean of 0.92 (±0.46)
mg/l. The nitrate concentration varied between 0.12 and
3.22 mg/l, with an average of 1.64 (0.47) mg/l. The
results indicate a lack of seasonal variations in both
nitrogen compounds (table 15). The NO_3/NH_4 ratio varied
between 0.08 and 4.21, with a mean of 2.28 (±1.00). This
average ratio is often found in precipitation samples and
also, with the exception of one outlier of 0.08, the
observed variability is frequently recorded in precipitation
(Likens & Bormann 1974). The ammonium and nitrate concentrations show for the most part a rather poor relationship with the other chemical parameters (table 15). For
example nitrate shows a significant negative correlation
with the frequency of westerly winds and with sodium,
indicating that this nitrogen compound is mainly derived
from terrestrial sources. The highly significant
relationship of nitrate with the pH is striking, suggesting
that nitrate contributes considerably to the free acidity.
Because nitrate and especially ammonium levels could be
influenced by biochemical reactions in the rain collector,

the statistical analyses were also applied for the total
nitrogen (NH_4 + NO_3) concentrations. These concentrations
are significantly correlated with sulphate (r = 0.576;
α chosen as 0.05), suggesting that nitrogen and sulphate
in the precipitation originate from similar source(s)
or that chemical reactions of these elements in the
atmosphere are producing nitrogen and sulphur compounds.
So the relationship between NO_3 and NH_4 suggests the
presence of NH_4NO_3 and the relationship N - SO_4 suggests
the presence of ammonium sulphate in the atmosphere.

Sulphur. The sulphur cycle in the atmosphere involved
primarily H_2S, SO_2, SO_3 and various sulphates. In unpolluted
air only three sulphur compounds SO_2 and H_2S as gases and
$SO_4{}^{2-}$ in aerosols are important. In the atmosphere H_2S is
converted rapidly into SO_2, which in turn is oxidised to
SO_3. The latter dissolves in water vapour and forms
sulphuric acid, which may react further to form sulphate
salts like $(NH_4)_2SO_4$. Both H_2SO_4 and sulphate salts exist
in the atmosphere as aerosols, which are removed by
precipitation and to a lesser extent by gravitational
settling. Estimated on a global scale about 30 percent of
the total amount of the sulphur mobilization in the
atmosphere has an anthropogenic origin due to fossil fuel
combustion and other industrial processes (table 16;
Kellogg *et al* 1972; Friend 1973). The residence time of
SO_2 in the atmosphere has been estimated as ranging from
5 days to 2 weeks (Eriksson 1963).

In the bulk precipitation in the Haarts catchment,
the sulphate variability was rather high, ranging from a
maximum of 8.65 mg/l to a minimum of 1.65 mg/l, with an
average of 3.57 (±1.99) mg/l. The results show that the
concentration of sulphate was inversely related to the
amount of precipitation and frequency of westerly winds
(table 15), suggesting that anthropogenic or at least
terrestrial sources were very important. This is also
corroborated by the high correlation between sulphate and
calcium (r = 0.688), indicating a direct relationship
between these two components. Sulphate was also highly
positively correlated with potassium.

The extensive reviews by Eriksson (1959, 1960),
Gorham (1955, 1958, 1961) and the recent reports by Bolin
(1971) and Oden (1976) all indicate that most of the SO_2
and H_2S from natural or anthropogenic sources are oxidized
to sulphuric acid. Consequently, if the produced acid is
not neutralized by alkaline substances, principally calcium,
magnesium and ammonium, there will be an excess of acid in
the atmosphere (Granat 1972a; Cogbill & Likens 1974).
Therefore in many investigations a very high correlation
between sulphate and the pH could be established. In the
Haarts catchment, however, no such relationship could be
established, probably due to the rather important influence
of nitric and hydrochloric acid (see also next paragraph).

Table 16 Estimates rates of sulphur mobilization from biogenic sulphur omissions, fossil fuel combustion, industrial sources and volcanoes (Gorham 1976)

Source	Sulphur (10^6 metric t yr^{-1})	
	Friend (1973)	Kellogg et al.(1972)
Natural		
Sea spray[*]	44	43
Biogenic (sea)[*]	48 ⎫	
Biogenic (land)	58 ⎭	89
Volcanoes	2	0.7
	Total 152	133
Anthropogenic		
Fossil Fuel combustion	51 ⎫	
Non-fuel sources	14 ⎭	50
	Total 65	50

[*]Most marine emissions return directly to the sea

Table 17 Acids found in atmospheric precipitation

Acids	strength (pK_a)	Acids	Strength (pK_a)	Acids	Strength (pK_a)
HCl	strong	Fe^{3+}	⎫	H_2CO_3	6.35
HNO_3	strong	$FeOH^{2+}$	⎪ 2.2 to 8.3	H_2S	6.9
H_2SO_4	strong	$Fe(OH)_2^+$	⎪	HSO_3^-	7.2
H_2SO_3	1.9	$Fe(OH)_3$	⎭	$H_2PO_4^-$	7.2
HSO_4^-	1.9	Al^{3+}	⎫	$B(OH)_3$	9.0
HF	3.2	$AlOH^{2+}$	⎪ 4.9 to ~8.2	NH_4^+	9.3
RCOOH[*]	3 to 6	$Al(OH)_2^+$	⎪	HCN	9.4
phenols	4 to 9	$Al(OH)_3$	⎭	H_4SiO_4	9.9

[*]organic acids

83

2.1.3.4 Acidity

The increasing acidity of many ecosystems, especially in Scandinavia and the North Eastern United States due to the impact of acids by precipitation (Eriksson 1952a and b; Barret & Brodin 1955; Gambell & Fisher 1966; Likens et al 1972) is a subject which has recently received considerable attention. Theoretical explanations of causes and origins have also been proposed. For Western Europe it could be established that most acidity in precipitation can be explained by the formation of sulphuric and nitric acids produced by industrial activities (Granat 1972a).

An acid is an electrically neutral or charged group that can donate protons to a water system. The strength of the donating power determines the strength of the acid. If the protonization is made completely even at high concentrations the acid is strong, otherwise it is weak. The acids which might occur in precipitation are presented in table 17. In any aqueous solution there are two types of acidity that can be established, ie. free acidity and total acidity (Galloway et al 1976a, b). Free acidity (free H^+ ions) can be measured by a glass electrode and is normally characterised by the negative logarithm of the proton activity (pH). Its amount in precipitation is determined by the amounts of strong acids (table 17). The total acidity is the amount of free H^+ ions and the H^+ ions bound in weak acids (for example H_2CO_3 and organic acids) and in Brønsted acids (for example dissolved Al, dissolved Fe and NH_4^+) (table 17). The value can be determined by a titration, which is a quantitative addition of a strong base to a water system. Thus it will be clear that in an aqueous solution containing only strong acids, the free acidity equals the total acidity, but in a system which is a mixture of strong-, weak- and Brønsted acids the total acidity is higher than the free acidity.

The free acidity of unpolluted atmospheric precipitation is theoretically governed by the atmospheric partial pressure of CO_2. The following equilibrium reactions prevail in this $H_2O(l)-CO_2(g)$ system:

$$CO_2(g) \rightleftharpoons CO_2(aq) \qquad (1)$$

$$CO_2(aq) + H_2O(l) \rightleftharpoons H_2CO_3^x(aq) \qquad (2)$$

$$\text{or} \quad CO_2(g) + H_2O(l) \rightleftharpoons H_2CO_3^x(aq) \qquad (3)$$

$$H_2CO_3^x(aq) \rightleftharpoons H^+(aq) + HCO_3^-(aq) \qquad (4)$$

$$HCO_3^-(aq) \rightleftharpoons H^+(aq) + CO_3^{2-}(aq) \qquad (5)$$

$$H_2O(l) \rightleftharpoons H^+(aq) + OH^-(aq) \qquad (6)$$

in which $H_2CO_3^x(aq) \rightleftharpoons CO_2(aq) + H_2CO_3(aq)$.

The electrostatic balance gives:

$$(H^+) = (OH^-) + (HCO_3^-) + 2(CO_3^{2-})$$

and utilizing the first and second dissociation constants of H CO*, B (equilibrium constant for overall reaction (3)) (table 18) K_w, and ignoring the influence of the activity coefficients because of the very low ionic strength of the system, and substituting for (OH^-), (HCO_3^-) and (CO_3^{2-}) in the electro-neutrality equation gives further:

$$(H^+)^3 = K_w (H^+) + B K_1 (H^+) P_{CO_2} + 2 B K_1 K_2 P_{CO_2}$$

If P_{CO_2} is greater than 10^{-4} bars, then the first and last term of the right hand side in this equation are insignificant compared to $BK_1 (H^+) P_{CO_2}$

therefore $(H^+)^3 = B K_1 (H^+) P_{CO_2}$

or $(H^+) = B K_1 P_{CO_2}$

or $pH = \frac{1}{2}(pB + pK_1) + \frac{1}{2} p P_{CO_2}$

When the maximum and minimum temperatures observed (20^oC and 0^oC) and the values of the constants (table 18) are substituted in this equation, temperature alone will cause a seasonal variation of equilibrium pH as follows:

Winter (0^oC)   pH 5.54

Summer (20^oC)   pH 5.59

Next to the influence of the temperature, another seasonal effect has to be considered. Bischof (1960) and Machta (1972) showed a pronounced seasonal variability in the level of atmospheric CO_2, with a maximum in January of 350 ppm ($P_{CO_2} = 10^{-3.453}$) and a summer minimum of 300 ppm ($P_{CO_2} = 10^{-3.522}$). These data superimposed on the temperature influence result in:

Winter: 0^oC, $P_{CO_2} = 10^{-3.455} \rightarrow pH = 5.57$

Summer: 20^oC, $P_{CO_2} = 10^{-3.522} \rightarrow pH = 5.65$

These variations are within the error of a pH measurement with portable field pH-meters. Thus it can be concluded that water in an unpolluted atmosphere in equilibrium with mean CO2 pressure will attain a pH of 5.6 if no alkaline substances play a quantitatively important role.

However, in the investigated area the pH in precipitation samples varied from a maximum of 5.53 to a

85

Table 18   Equilibrium constants (of the reactions (3), (4) and (5) in the $H_2O(l)$ - $CO_2(g)$ system as a function of temperature

Temperature °C	Log B	Log $K_{a1}H_2CO_3$	Log $K_{a2}HCO_3$
0	-1.1151	-6.576	-10.626
1	-1.1317	-6.564	-10.611
2	-1.1480	-6.551	-10.596
3	-1.1642	-6.539	-10.582
4	-1.1801	-6.528	-10.568
5	-1.1957	-6.516	-10.554
6	-1.2112	-6.505	-10.540
7	-1.2264	-6.495	-10.527
8	-1.2415	-6.484	-10.514
9	-1.2563	-6.474	-10.501
10	-1.2709	-6.464	-10.488
11	-1.2853	-6.455	-10.476
12	-1.2994	-6.445	-10.464
13	-1.3134	-6.436	-10.452
14	-1.3272	-6.428	-10.441
15	-1.3407	-6.419	-10.429
16	-1.3541	-6.411	-10.418
17	-1.3673	-6.404	-10.408
18	-1.3803	-6.396	-10.397
19	-1.3930	-6.389	-10.387
20	-1.4056	-6.382	-10.377
21	-1.4180	-6.375	-10.367
22	-1.4302	-6.369	-10.357
23	-1.4423	-6.363	-10.348
24	-1.4541	-6.357	-10.339
25	-1.4658	-6.351	-10.330

minimum of 4.00, with an average of 4.44. No pronounced seasonal pattern is evident, and the pH is significantly negatively correlated with nitrate (r = -0.61, α < 0.05). Thus the precipitation is acid (pH < 5.6, mostly pH < 5.0) being a mixture of strong-, weak- and Brønsted acids, but because (1) the weak and Brønsted acids are in low concentrations, (2) their dissociation constants are much less than unity (very weak acids) and (3) because most are only slightly disociated at pH < 5.6, they contribute primarily to the total acidity and only negligibly to the free acidity (Galloway et al, 1976 a, b).

Unfortunately no total acidity determinations were made (titration with NaOH to pH = 9), so that this quanitity has to be indirectly approximated. For this it is assumed that the amount of particulate matter, such as clay minerals, is negligible and thus does not contribute to the total and free acidity. The influence of organic acids on the total acidity is relatively unimportant, due to the fact that only trace amounts were found in (bulk) precipitation samples. Further, because the pK values are about 4 (table 17) their maximum contribution to the free acidity of acid precipitation would be at most a very few percents (Galloway et al 1976 a, b).

The concentration of dissolved Al in the precipitation samples varied from 0 to 3 μmole/l, therefore its contribution to the total acidity ranged from 0 to 10 μeq/l and it has no effect on the free acidity (table 17). Also the effect of the dissolved Fe concentrations has to be considered. The dissolved Fe levels varied from 0 to 1 μmole/l; Fe is dominantly present as the $FeOH^{2+}$ or $Fe(OH)_2^+$ complexes in these acid precipitation samples. At pH 9 soluble ferric iron is present for about 80 percent as the $Fe(OH)_4^-$ complex and for about 20 percent as the soluble $Fe(OH)_3^0$ complex or it is already precipitated as $Fe(OH)_3(s)$. These changes will contribute less than 3 μeq/l to the total acidity. Its effect on the free acidity is negligible because it is less than 1 μeq/l. The contribution of dissolved silica and dissolved manganese, both with concentration levels of much less than 1 μmole/l on the total and free acidity can be ignored.

The contribution of the carbonic acid to the total acidity can be easily estimated for the samples with a pH < 5 from the partial CO_2 pressures. When pH < 5, the total carbonic acid concentration is completely determined by the $H_2CO_3^*$ concentration. This concentration varies between 10.7 and 12.9 μmole/l and it will contribute 21 to 26 μeq/l to the total acidity. For these acid precipitation samples, carbonic acid does not contribute to the free acidity. However, if pH >5 there will be a small effect of the bicarbonate concentration on the free and total acidity.

The effect of ammonium on the total acidity can be rather important. The ammonium concentration levels in the precipitation samples ranged between 14 to 91 μeq/l with an average of 51 μeq/l. In a titration to pH 9 about one third of the NH_4^+ will be converted to NH_3 (table 17). Thus it can be concluded that the contribution of ammonium to the total acidity of the Haarts precipitation samples will be in the range from 5 to 30 μeq/l. Ammonium does not contribute to the free acidity due to the low pH of the precipitation samples.

From the above it can be concluded that for the Haarts catchment precipitation, the free acidity varied between 3 μeq/l (pH = 5.53) and 100 μeq/l (pH = 4.00) with an average of 37 μeq/l (pH = 4.43). This free acidity is almost completely attributed to the excess amounts of sulphuric, nitric and hydrochloric acids, with only a very small contribution of weak acids such as organic acids and sometimes of dissolved iron and carbonic acid (the latter if the pH >5). The total acidity, however, can be of an order of magnitude higher, up to 160 μeq/l, mainly due to the high contributions of ammonium, carbonic acid and perhaps organic acids. The influences of dissolved Al, Fe, Mn and Si on this parameter are very limited.

The relationship between the free acidity and the total amount of acid or alkaline substances was investigated by Granat (1972a). A stoichiometric relationship exists for the major chemical ions in precipitation (fig. 18). Certain cations come from sea-salt particles and are effectively neutralized by corresponding anions. Therefore sodium, because it is more or less completely of maritime origin,can be used as a basis for tracing this sea-salt effect, and the ratio of sodium to the other ions in sea-water can be used to predict the equivalent amounts of Ca, Mg, K and SO_4. Because the chloride/sodium ratios for the Haarts precipitation samples are almost always above 1.80 (table 14), unlike in Grant's model, chloride was treated in the same way as the other 4 ions. Only when the chloride/sodium ratio was less than that of seawater, was chloride used as the basis. The remaining ions, commonly called "excess" ions, are assumed to originate from terrestrial sources. Each equivalent of the excess cations was originally combined with one equivalent of carbonate and each equivalent of excess anion was associated with hydrogen loss. Thus a certain amount of these excess ions neutralised each other (fig. 18). Equal portions of the remaining sulphate, nitrate and chloride equivalents are associated with hydrogen ions. The sum of these hydrogen ions equivalents gives a predicted hydrogen ion concentration or pH that can be compared to the measured pH. Thus the model of Granat, also modified slightly to incorporate new data on the concentrations of ions in seawater (Riley & Skirrow 1975) shows that the amount of available acid was equal to:

$$a = 2\left(m_{SO_4} - m_{Na}\frac{28.4}{470}\right) + m_{NO_3} + \left(m_{Cl} - m_{Na}\frac{547}{470}\right) - m_{NH_4}$$

in which m denotes molarities, or for the case that Cl/Na ratio was less than that of seawater:

$$a = 2\left(m_{SO_4} - m_{Cl}\frac{28.4}{547}\right) + m_{NO_3} - m_{NH_4}$$

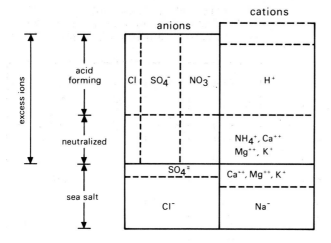

Figure 18 Theoretical relationship between major
chemical ions in precipitation (from Cogbill
& Likens 1974). Any lack of equivalence
between anions and cations can be envisaged
to exist as a misestimate of H^+ concentration.

and the amount of available carbonate:

$$b = \tfrac{1}{2}(m_K - m_{Na} \frac{10}{470}) + (m_{Mg} - m_{Na} \frac{54}{470}) + (m_{Ca} - m_{Na} \frac{10.25}{470})$$

or for the case that the Cl/Na content was less than that
of seawater:

$$b = \tfrac{1}{2}(m_K - m_{Cl} \frac{10}{547}) + (m_{Mg} - m_{Cl} \frac{54}{547}) + (m_{Ca} - m_{Cl}\frac{10.25}{547})$$

$$+ \tfrac{1}{2}(m_{Na} - m_{Cl} \frac{470}{547})$$

Consequently the amount of acid not neutralized is:

$$e = a - 2b \quad\quad\quad \text{and} \quad\quad\quad pH = - \log e$$

  All the samples with a pH < 5.0 and a reasonable
cation-anion balance were used in the calculation of the
theoretical hydrogen ion content based on the equations
given above (table 19). The calculated pH of these samples
shows a strong significant relation ($r = 0.649$, $\alpha < 0.05$)
with the measured pH, suggesting that the modified model
of Granat (1972a) is useful for acid precipitation. Table
19 also gives the contributions of the "excess" anions to
the free acidity and it appears that "excess" sulphate
contributed between 4.9 and 78.4 percent of the free
acidity in the various rainfall samples, with a mean of
47.8 percent. The contribution to the free acidity of

Table 19   Measured and calculated pH values and the
           contribution of the "excess" anions to the free
           acidity for precipitation samples with pH < 5

pH measured	pH calculated	% free acidity associated with		
		SO_4^{2-}	NO_3^-	Cl^-
4.70	4.67	78.4	22.6	-
4.95	4.89	51.9	48.1	-
4.50	4.15	58.8	17.6	23.6
4.22	4.20	51.5	15.6	33.0
4.92	4.95	67.7	17.7	14.5
4.70	5.06	38.5	39.9	21.6
4.41	4.41	57.0	27.0	16.0
4.41	4.43	57.6	30.3	12.1
4.46	4.60	4.9	11.5	83.6
4.50	5.00	23.9	50.2	25.8
4.25	4.64	35.6	64.4	-

nitrate ranged from 11.5 to 64.4 percent, with a average
of 31.4 percent. "Excess" chloride varied between 0 and
83.6 percent, with a mean of 20.9 percent. These data
also support the opinion that the source of the acidity
originates from anthropogenic activities.

Another parameter for monthly precipitation samples
in relation to acidity has also been used. Eriksson
(1969) introduced the concept of "excess acids". This
quantity is calculated by subtracting the sum of the
alkalinities for those months with pH > 5.6 from the sum
of acids in monthly samples with pH < 5.6. The "excess
acids" equal the amount of strong acids minus the
alkalinity. This parameter leads to a fairly large underestimation of the fallout of acids on a regional scale,
due to the local contamination by dust, lime and ashes etc.,
and therefore gives alkalinities which are too high (Oden
1976). According to this author, the total acidifying
effect of precipitation can only be established if a proper
state of reference is introduced. With respect to the
acidity problem, such a natural reference is the pH of
non-polluted precipitation, which for Western Europe
probably will be pH ∿ 7 to 7.5 and not pH = 5.6 due to the
influence of natural alkaline substances. Since the
natural pH level is not known with any real certainty, the
total acidifying effect of precipitation is equally
uncertain. What is in fact measured is not the level of the
acidifying effect, but variations in it (Oden 1976).

2.2 Spring- and soilwater chemistry

In this section the chemistry of the soil- and springwater in relation to the water-rock interactions will be discussed. Soil waters were obtained with the help of porous cup soil water samplers (Parizek & Lane 1970; Wood 1973). At the ground surface these were covered with plastic sheets to prevent the direct downward flow of precipitation along the tube-wall. Due to the high gravel and stone content of the soils in the Haarts catchment the porous cup soil water samplers could only be used to a depth of approximately 90-100 cm. With this method soil water can be collected which is bound to the soil particles with a tension of less than 0.8 bar (pF < 2.9). During the summer periods no soil water samples could be obtained by this method. Soil water from various depths was also gathered in PVC-cylinders, perforated at their upper surfaces, which served as drains to polyethene bottles. This method often failed to obtain clear and unpolluted soil water samples, because soil material was also collected into the polyethene bottles. These samples were discarded because they can represent artificial "equilibrium" conditions between this soil material and the collected soil solution which perhaps do not exist in the soil. Springwater samples were collected throughout the year.

Springwater composition indicates that carbon dioxide is the driving force for weathering in the Haarts catchment. The main cations Mg, Na, Ca, K are all strongly positively related with the bicarbonate content (table 20), while no relationship exists with other proton donors like chloride, sulphate and nitrate. No direct measurements of the $CO_2(g)$ concentrations in the soil atmosphere were done, but springwater P_{CO_2} was calculated from the pH and alkalinity. The calculated partial CO_2 pressures show a clear seasonal trend (fig. 19). Seasonal variations in CO_2 production in the soil atmosphere are best considered in relation to the major climatological factors, as expressed by soil temperature and soil moisture. The evolution of CO_2 is positively correlated with the temperature. This is for instance clearly represented by the marked rise in CO_2 production during springtime when soil temperature increases strongly. Also the effect of soil moisture often plays an important role in the CO_2 concentrations, although this factor seems to be secondary to the soil temperature (Edwards 1975). Soil moisture contents, especially of the deeper soil horizons, show only slight variations (fig. 20), moreover, the high gravel and stone content of the soil interferes with precise measurements. Thus it can be concluded that soil temperature is probably the more important indirect factor which influenced partial CO_2 pressures. From field measurement, hydrograph analyses, temperature variations of the springwaters throughout the year and lithological data it was concluded that all the springwaters are coming from the saprolites and perhaps the uppermost part of the hard rock in the Haarts catchment.

Table 20 Matrix of linear correlation coefficients[*] of dissolved species, electrical conductivity and temperature for springwaters (n = 90)

	pH	K	Na	Ca	Mg	H_4SiO_4	NO_3	Cl	SO_4	HCO_3	EC_{25}	Temp
pH	1.00	0.43	0.42	0.51	0.56	0.51	-0.51	0.01	0.13	0.62	0.53	0.41
K		1.00	0.78	0.77	0.84	0.10	-0.31	-0.18	0.10	0.80	0.80	0.61
Na			1.00	0.82	0.88	0.14	-0.17	-0.29	0.24	0.82	0.87	0.72
Ca				1.00	0.90	0.14	-0.20	-0.28	0.17	0.88	0.91	0.80
Mg					1.00	0.21	-0.36	-0.30	0.16	0.97	0.96	0.76
H_4SiO_4						1.00	-0.48	-0.14	0.14	0.26	0.12	0.17
NO_3							1.00	0.20	0.15	-0.46	-0.17	-0.32
Cl								1.00	-0.07	-0.33	-0.25	-0.27
SO_4									1.00	0.15	0.25	0.00
HCO_3										1.00	0.96	0.80
EC_{25}											1.00	0.77
Temp.												1.00

[*] $|r| \geq 0.175$ $\alpha = 0.10$; $|r| \geq 0.208$ $\alpha = 0.05$; $|r| \geq 0.245$ $\alpha = 0.02$; $|r| \geq 0.270$ $\alpha = 0.01$.

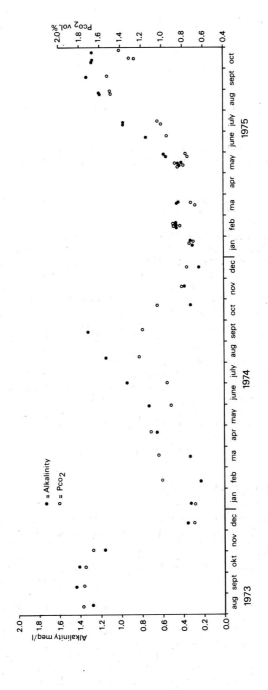

Figure 19   Seasonal fluctuations in the partial CO_2 pressures and the alkalinities in springwater

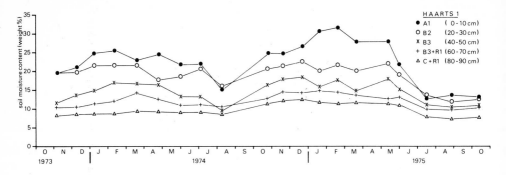

Figure 20   Fluctuations in the soil moisture content of the soils on the steep lower valley slopes

Therefore the springwaters represent soil environmental conditions. This means that CO_2 is added to the water in the soils and variation in its value must be due to the partial CO_2 pressures in the soil atmosphere and to the infiltration conditions. If infiltration of precipitation water is rapid, saturation with CO_2 occurs first and weathering is limited due to the short contact time. This results in relatively low alkalinities (winter periods). If infiltration is slow, relatively consideraly alteration reactions may take place in the soils and water is continuously fed with CO_2 (partly) replacing the amount that is consumed. It is clear that the latter type of water infiltration, if the influence of partial CO_2 pressure variations is not taken into account, will result in higher alkalinities and total carbonates and ultimately more cations. This effect of difference in infiltration rate of the precipitation water on the chemical composition of the spring- and streamwaters is obvious under various hydrological conditions in the Haarts catchment (Imeson & Verstraten 1979), although the influence of the partial CO_2 pressures plays an important role.

2.2.1   Springwater chemistry

If precipitation water enters the soils in the catchment qualitative and quantitative chemical changes in its composition occur if it passes through the system. Water enters the system mainly as a dilute solution of sulphuric, nitric and hydrochloric acid (pH ∿4.4) but leaves the ecosystem containing primarily bicarbonates, neutral sulphates and to a lesser extent neutral nitrates (pH 6-7). The mean ionic strength of the streamwater is 1.1 meq/1 (Verstraten 1977), while that of the precipitation is 0.4 meq/1. From data given in table 3J it can be concluded that the average evapotranspiration loss is about 50%. Because the catchment is watertight, this evapotranspiration leads to an ionic strength of about 0.8 meq/1, suggesting the important contribution of this factor on the "nutrient" budget. However, the use of such a con-

94

centration factor in purely input-output studies is
somewhat misleading, because in the ecosystem (biomass +
saprolite) important internal chemical and biological
reactions occur, resulting in changes in concentrations
of the spring- and streamwaters, but also in the
proportions of the dissolved species as well.

The concentration levels of magnesium, calcium,
sodium and potassium in the springwaters are all
significantly positively correlated with the alkalinity
(r = 0.96; 0.87 and 0.80; table 20) indicating the
importance of carbondioxide as weathering agent. These
cations are all strongly positively correlated to each
other (table 20), suggesting a common origin and one
mechanism which releases these nutrients from the rocks
and weathering materials. The relationship between
aqueous silica levels and the alkalinity is only slightly
positive (r = 0.26), indicating a multiple source of this
constituent. McKeague & Cline (1963a, b) concluded that
no relationship exists between the orthosilicic acid
levels in soil solutions and soil clay mineralogy, and
that the behaviour of aqueous silica in soils is controlled
by a pH-dependent adsorption mechanism, resulting in higher
aqueous silica concentrations in soil solutions at lower
pH levels, due to less adsorption of silica by sesquioxides
(Beckwith & Reeve 1963). Next to sorption-desorption by
soil components of soluble silica also leaching of orthosilicic acid from the soil and plant uptake are important
in determining the aqueous silica concentration in soils.
However, from the correlation coefficients of aqueous
silica with alkalinity (r = 0.26) and pH (r = 0.51) it was
preliminarily concluded that simple dissolution-precipitation
reactions of rock and soil minerals are the dominant factor
for the aqueous silica concentration levels, although
other sources cannot be excluded.

For the main spring in the Haarts catchment,
regression equations of the main cations, aqueous silica
and the alkalinity are given in fig. 21. The points
representing magnesium, sodium and calcium concentrations
do not show any tendency to level out at higher alkalinity
values, which might be expected when (partial) equilibrium
conditions prevail in the weathering system. Only the
points for aqueous silica and potassium do not keep pace
with a linear rise at higher alkalinities. From the
regression equations the molar ratios of magnesium, calcium,
sodium, potassium, aqueous silica and bicarbonate were
calculated; they are 18.2:7.2:4.5:5.1:1.0:1.9:54.5.
These molar ratios indicate, if those in the parent rock
are taken into account (see B 3.3), incongruent dissolution
of the parent rocks, resulting in new formation of one or
more solid phases. These phases could consist of amorphous
materials, clay minerals and crystalline pedogenic oxides.
From the results of the various extraction techniques
(see A II 6 and appendix II) it can be concluded that
amorphous Al and Si compounds are unimportant in the soils
and weathering materials in the Haarts catchment. Therefore, it is assumed that these compounds play at most a

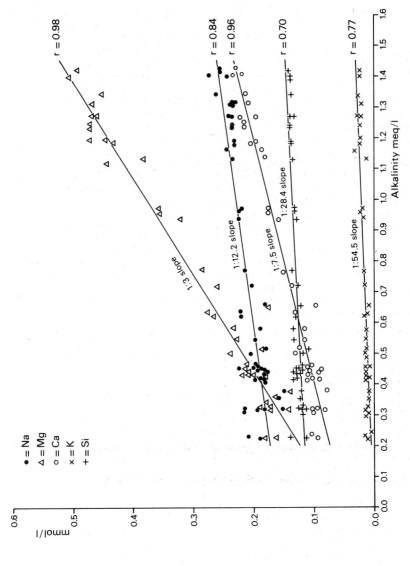

Figure 21 Linear regression of the main cations and aqueous silica with the alkalinity (spring 2)

96

very subordinate role in the springwater chemistry, while clay minerals play an important role. The very low aluminium concentrations in the spring waters confirm the incongruent dissolution processes. The aluminium concentrations are below our detection limit (9 $\mu g/l$). On the base of these data it is obvious that springwaters are clearly undersaturated with respect to albite, muscovite and illite. However, in the "absence" of aluminium it is not possible to determine the state of saturation with respect to secondary minerals such as kaolinite, vermiculite and smectite. An alternative approach is to assume that aluminium is an inert component (Thompson 1955) and to examine the stability relationship in terms of activities of Mg^{2+}, H^+, K^+ and $H_4SiO_4^0$. If the chemical composition of the springwaters is plotted in stability diagrams, the composition of the samples representing low flow conditions, especially in the summer periods, are all close to the kaolinite-vermiculite-smectite boundary (fig. 22a). The position of the waters more to the centre of the kaolinite stability field appears to be a result of mixing of waters of different origin or the result of open system conditions, representing stages along the various weathering paths depending on flow and weathering rates (see B 3.3.3).

In order to detect whether partial equilibria conditions prevail in the weathering system, the departure from equilibrium in a given partial system - the disequilibrium index $D_{\alpha-\beta}$ - has been calculated (Paces 1973). This parameter is defined for a chemical reaction:

$$a A + b B \rightleftarrows c C + d D$$

as $\quad D_{\alpha-\beta} \quad = \log_{10}(\frac{Q}{K}) = \Delta G/2.303 \; RT$

with $\quad Q = \dfrac{(C)^c \; (D)^e}{(A)^a \; (B)^b}$ (= activity product)

where (A), (B), (C), (D) are activities of the participating compounds A, B, C, D; a, b, c, d are stoichiometric coefficients;

$\alpha-\beta$ is a symbolic description of the left hand and right hand sides of the reaction;

K is the equilibrium constant of the reaction (at equilibrium $K = Q$ and $D_{\alpha-\beta} = 0$);

Δ G is the free energy of reaction at a given state of development of the system;

R is the gas constant;

T is the absolute temperature of the system.

If A represents a solid phase and B, C and D dissolved compounds, a positive $D_{\alpha-\beta}$ indicates supersaturation and a

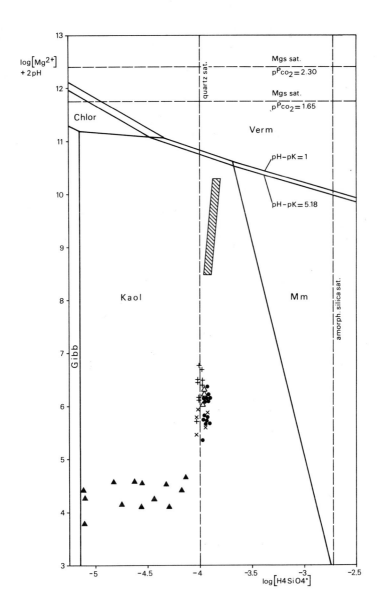

Figure 22a   Stability fields of gibbsite, kaolinite, Mg-
beidellite, clinochlore and vermiculite in the
presence of amorphous ferric iron (298.15^{0}K,
1 bar pressure). Symbols represent composition
of spring- and soil waters.

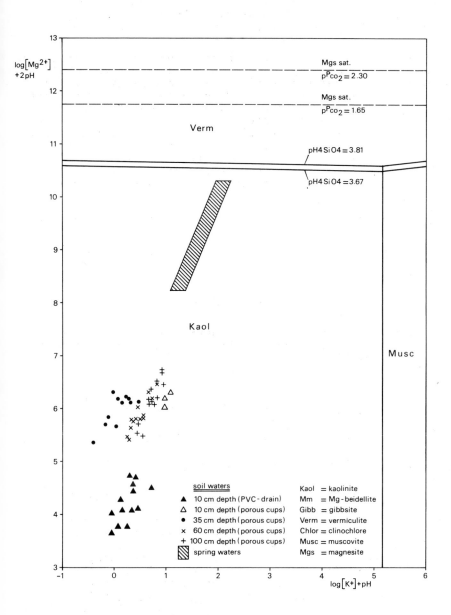

Figure 22b   Stability fields of kaolinite, muscovite and vermiculite in the presence of amorphous ferric iron (298.15^{0}K, 1 bar pressure). Symbols represent composition of spring- and soil waters.

negative value undersaturation of the aqueous solution with respect to solid phase A. If A and C represent solid phases and B and D dissolved species the positive values indicate stability of solid phase A and negative values stability of solid phase C.

Due to the undetectable aluminium and iron concentrations in the springwaters only incongruent transformations could be evaluated, assuming aluminium and iron conservation. As states before, this approach produces values indicating that albite and muscovite (illite) will never be stable solid phases in the weathering system. From the springwater compositions plotted in fig. 22a it is also clear that primary chlorite (clinochlore) will never be a stable phase. The disequilibrium indices of kaolinite with respect to vermiculite and Mg-beidellite were calculated, by con-sidering the reactions: Vermiculite/Kaolinite + amorphous ferric iron:

$$K_{0.1}Mg_{6.69}Fe^{III}_{0.59}Al_{1.61}Si_{5.98}O_{20}(OH)_4(s) + 13.48H^+ + 2.495H_2O \rightleftharpoons$$

$$0.805Al_2Si_2O_5(OH)_4(s) + 0.59Fe(OH)_3(s) + 6.69Mg^{2+} + 0.1K^+ +$$

$$4.37H_4SiO^0_4$$

$$D_{Verm-Kaol} = 54.96 + 13.48pH - 6.69pMg - 0.1pK - 4.37pH_4SiO^0_4$$

where pX is the negative decadic logarithm of the activity of dissolved species X and amorphous ferric iron governs the ferric iron solubility.

Mg-beidellite/Kaolinite:

$$Mg_{0.167}Al_{2.33}Si_{3.67}O_{10}(OH)_2(s) + 0.33H^+ + 3.845H_2O \rightleftharpoons$$

$$1.165Al_2Si_2O_5(OH)_4(s) + 0.167Mg^{2+} + 1.34H_4SiO^0_4$$

$$D_{Mg-Beid-Kaol} = 3.18 + 0.333pH - 0.167pMg - 1.34pH_4SiO^0_4$$

The $D_{Verm-Kaol}$ and $D_{Mg-Beid-Kaol}$ values are all negative

(fig 23 and 24) indicating the stability of kaolinite at first sight. However, when vermiculite and smectites, with a different chemical composition and consequently a slightly deviating free enthalpy of formation value are used, the disequilibrium indices of these clay minerals can approach to zero. These calculations were performed with various smectites, and less negative $D_{smectite/Kaolinite}$ values were obtained. Due to the lack of thermochemical data for vermiculites no additional information in absolute sense could be obtained for disequilibrium indices vermiculite-kaolinite. The smectite content in the soil and weathering materials in the Haarts catchment is very low (see B 1.2) and no additional X-ray diffraction analyses could be carried out in order to establish the smectite species.

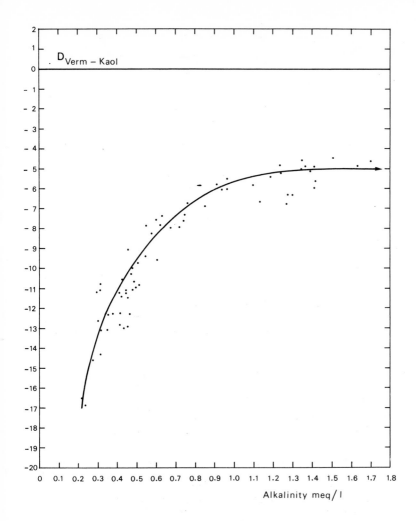

Figure 23   Disequilibrium indices of reaction vermiculite ⇌
            kaolinite + solution ($D_{Verm-Kaol}$) as a function
            of the alkalinity (springwaters)

Therefore the values of the alternative smectite/kaolinite
disequilibrium indices are somewhat speculative and only
the $D_{Mg-Beid-Kaol}$ are presented here (fig. 24). Also for
the use of $D_{Verm-Kaol}$ in absolute sense, serious limitations
exist. The vermiculite used in the departure from
equilibrium evaluation is a macroscopic vermiculite. Real
clay vermiculites which occur in soils have definitely
different chemical compositions (Weaver & Pollard 1975)
and consequently different disequilibrium indices. Exact
knowledge on the chemical composition of the vermiculite
and smectite in the Haarts catchment is lacking.
Consequently it seems to be a more reliable procedure to

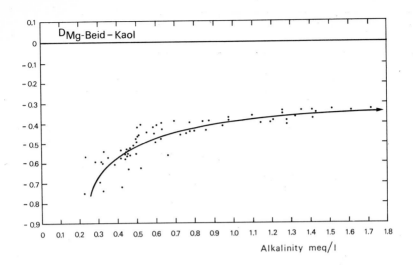

Figure 24   Disequilibrium indices of reaction
             Mg-beidellite ⇌ kaolinite + solution
             $(D_{Mg-Beid-Kaol})$ as a function of the alkalinity
             (springwaters)

evaluate the curves of the log (Mg^{2+})+2pH, log(K^+)+pH
and the log$(H_4SiO_4^0)$ values plotted with the alkalinity
(fig. 25) and those of the $D_{Verm-Kaol}$ and $D_{Mg-Beid-Kaol}$
values also plotted with the alkalinity (fig. 23 and 24)
in order to trace whether partial equilibrium conditions
occur.

     Levelling tendencies in the curves of the log(Mg^{2+}) +
2pH and log(K^+)+pH values are evident at higher alkalinities
(fig. 25). The orthosilicic acid concentrations only very
slightly fluctuate in the springwaters (fig. 21),
consequently no log$(H_4SiO_4^0)$ curve is given. However from
fig. 21 some tendency of levelling in the aqueous silica
concentrations at higher alkalinities can be established.
All these parameters show similar levelling at identical
conditions ie. higher alkalinities, therefore it is
preliminarily concluded that partial equilibria between
vermiculite and kaolinite and/or smectite and kaolinite
exist. Even if disequilibrium indices for other smectites
and vermiculites are used, similar curves as shown in fig.
23 and 24 are obtained, indicating the partial equilibrium
state(s). From the mineralogical investigations (see
B 1.2) it seems reasonable to conclude that especially the
vermiculite-kaolinite partial equilibrium is effective.

     The straight forward increase of the various cations,
especially magnesium, calcium and sodium with increasing
alkalinities (fig. 21) also suggests a rather simple
weathering mechanism without new formation of a complex
clay mineral assemblage. Such an increase in these
dissolved species concentration levels points to the

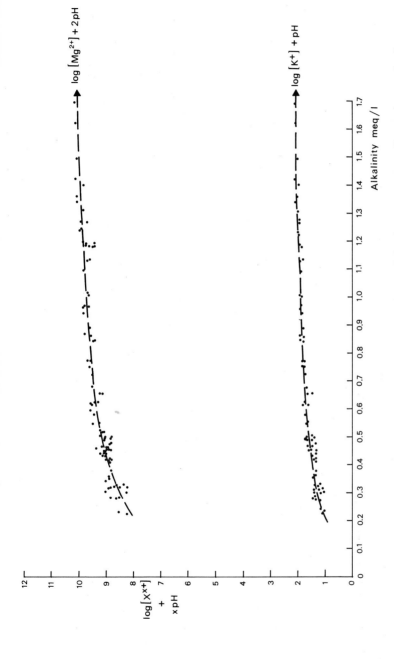

Figure 25   Activity ratios of magnesium, potassium and hydrogen as a function of the alkalinity (springwaters)

formation of kaolinite, if the aqueous silica activities are taken into account (kaolinitization stage). At alkalinities higher than about 1.2 meq/l the ultimate stage in this weathering system is reached, resulting in a partial equilibrium between vermiculite (smectite) and kaolinite (vermiculitization(smectitization)-kaolinitization stage).

An attempt has been made to determine from which minerals the various dissolved species in the springwaters originate in order to test the conclusions concerning the incongruent weathering reactions (Garrels & Mackenzie 1967). First the above-mentioned kaolinitization stage is discussed. As a representative for the end of this stage an average springwater composition was taken (table 21). The meteorological input is substracted from the spring-water solutes to determine the materials derived from the rock. The Na^+, HCO_3^- and H_4SiO_4 are allowed to react with kaolinite to form albite, using up all of the sodium. Next all of the K^+ and enough Mg^{2+}, HCO_3^- and H_4SiO_4 reacts with kaolinite to make K-mica (phengite). Then chlorite and aqueous silica are created by combining the rest of the Mg^{2+} with HCO_3^- and kaolinite. The remaining Ca^{2+} is used up by apatite precipitation. The mass balance, also for bicarbonate, gives good results in this kaolinitization stage (table 21). Only aqueous silica gives some trouble. At alkalinities above ca. 1.2 meq/l a partial equilibrium between kaolinite and vermiculite (smectite) exists, so these clay minerals will be formed. This vermiculitization (smectitization)-kaolinitization stage was also evaluated (table 22). From the springwater composition (at high alkalinities) the meteorologic input and the composition at the end of the kaolinitization stage were subtracted. Next, kaolinite with various dissolved species is used to form albite and K-mica. Also apatite is formed in this stage. No results could be obtained by the reverse reaction of chlorite from kaolinite and vermiculite with magnesium, because no exact knowledge exists on the composition of vermiculite (table 22). However, the unknown coefficients from the reaction have to be low, suggesting little kaolinite and vermiculite formation from Mg-(Fe)chlorite. The same is true for the kaolinite and smectite reaction (table 22).

From the results given in tables 21 and 22, it can be concluded that most of the rock is affected in the kaolinitization stage and that in the next vermiculitization (smectitization)-kaolinitization stage clay minerals are formed to a much lesser extent (see also B 4.3). K-mica seems to be only slightly attacked by chemical weathering in both stages. On the other hand, X-ray investigations and chemical analyses of the soil and weathering materials indicate that this mineral is transformed by weathering. Especially a mass ratio of destroyed albite/K-mica of about 10 seems to be somewhat too high if the molar ratios in the parent rocks and saprolites are considered. A selective fixation of potassium in the weathering environment by adsorption and/or uptake by plant roots from the

Table 21 Kaolinization stage to an alkalinity of 1.1 meq/l

Reaction (coefficient x 10³)	Na⁺	K⁺	Mg²⁺	Ca²⁺	$H_4SiO_4^0$	mineral altered and (product)
	mmoles/l					mmoles/l
Initial concentrations springwaters	0.24	0.025	0.44	0.20	0.135	
Minus concentrations atmospheric precipitation	0.21	0.02	0.43	0.18	0.135	
change kaolinite back into albite						
$0.105Al_2Si_2O_5(OH)_4 + 0.21Na^+ + 0.42H_4SiO_4 + 0.21HCO_3^-$ $\rightarrow 0.21NaAlSi_3O_8 + 0.21CO_2 + 1.15H_2O$	–	0.02	0.43	0.18	–0.285	0.21 albite (0.105 kaolinite)
change kaolinite back into K-mica (phengite)						
$0.02Al_2Si_2O_5(OH)_4 + 0.02K^+ + 0.01Mg^{2+} + 0.03H_4SiO_4 + 0.04HCO_3^-$ $\rightarrow 0.02\,K\,Mg_{0.5}Al_2Si_{3.5}O_{10}(OH)_2 + 0.04CO_2 + 0.1H_2O$	–	–	0.42	0.18	–0.31	0.02 K-mica (0.02 kaolinite)
precipitate apatite						
$0.18Ca^{2+} + 0.108PO_4^{3-} + 0.036OH^-$ $\rightarrow 0.036Ca_5(PO_4)_3OH$	–	–	0.42	–	–0.31	0.036 apatite
a. change kaolinite and goethite back into Mg-Fe^{II}-chlorite						
$0.315Al_2Si_2O_5(OH)_4 + 0.33\,FeOOH + 0.42Mg^{2+} + 0.84HCO_3^-$ $\rightarrow 0.175Mg_{2.4}Fe_{1.9}Al_{3.6}Si_{2.15}O_{10}(OH)_8 + 0.255H_4SiO_4 + 0.84CO_2 + 0.05H_2O + 0.08O_2$	–	–	–	–	–0.05	0.175 Mg-Fe^{II}-chlorite (0.315 kaolinite) (0.33 goethite)
b 1. change kaolinite back into Mg-chlorite						
$0.195Al_2Si_2O_5(OH)_4 + 0.42Mg^{2+} + 0.84HCO_3^-$ $\rightarrow 0.10Mg_{4.1}Al_{3.8}Si_{2.1}O_{10}(OH)_8 + 0.175H_4SiO_4 + 0.84CO_2 + 0.05H_2O$	–	–	–	–	–0.13	0.10 Mg-chlorite (0.195 kaolinite)
b 2. change goethite back into siderite	–	–	–	–	–0.13	siderite (goethite)

105

Table 22 Vermiculitization (smectitization)-kaolinitization stage to an alkalinity of 1.4 meq/l

Reactions (coefficients x 10^3)	Na^+	K^+	Mg^{2+}	Ca^{2+}	$H_4SiO_4^0$	mineral altered and (product)
			mmoles/l			mmoles/l
Initial concentrations springwaters	0.26	0.029	0.49	0.24	0.14	
minus concentrations atmospheric precipitation	0.23	0.024	0.48	0.22	0.14	
minus kaolinitization stage	0.02	0.004	0.05	0.04	0.05	
change kaolinite back into albite						
$0.01Al_2Si_2O_5(OH)_4 + 0.02Na^+ + 0.04H_4SiO_4 + 0.02HCO_3^-$	–	0.004	0.05	0.04	0.01	0.02 albite
$\rightarrow 0.02NaAlSi_3O_8 + 0.02CO_2 + 0.11H_2O$						(0.01 kaolinite)
change kaolinite back into K-mica (phengite)						
$0.004Al_2Si_2O_5(OH)_4 + 0.004K^+ + 0.002Mg^{2+} + 0.006H_4SiO_4 + 0.008HCO_3^-$		–	0.05	0.04	–	0.004 K-mica
$\rightarrow 0.004\ K\ Mg_{0.5}Al_2Si_3O_{10}(OH)_2 + 0.008CO_2 + 0.02H_2O$						(0.004 kaolinite)
precipitate apatite						
$0.04Ca + 0.024PO_4^{3-} + 0.008OH^-$			0.05	–	–	0.008 apatite
$\rightarrow 0.008Ca_5(PO_4)_3OH$						
a. change vermiculite and kaolinite back into chlorite						
p Kaolinite + q Vermiculite + r Goethite . $0.05Mg^{2+} + 0.10HCO_3^-$	–	–	–	–	?	chlorite
$\rightarrow sMg\text{-}(Fe)Chlorite + t\ H_4SiO_4 + 0.10CO_2 + u\ H_2O + v\ O_2$						(kaolinite) (vermiculite)
b. change smectite and kaolinite back into chlorite						
p Kaolinite + q Smectite + r Goethite + $0.05Mg^{2+} + 0.10\ HCO_3^-$	–	–	–	–	?	chlorite
$\rightarrow sMg\text{-}(Fe)Chlorite + t\ H_4SiO_4 + 0.10CO_2 + u\ H_2O + v\ O_2$						(kaolinite) (smectite)

earlier, the set of data of the soil waters is rather limited because the collection methods failed in drier periods.

The soil waters show several characteristics which are related with the soil environment, ie. in addition to those described in the previous section also the influence of soluble organic matter compounds and adsorption/desorption processes, especially in the upper parts of the soils. Fluctuations in the composition for the soil waters collected with porous cups were small, suggesting steady state conditions in the soil system. The solute levels in the soil waters of the surface horizon are higher than those from underlying horizons (fig. 26). These levels gradually decrease towards the B3 + R1 horizons (90 - 100 cm depth). Surface horizon soil waters contain much higher concentrations of aluminium and iron levels. They decrease particularly rapidly going from the surface horizon (10 cm depth) to the B horizon (35 cm depth). At lower depth (60 and 100 cm) aluminium and iron are not detectable in the soil waters. The only reasonable explanation for this behaviour, even if the slightly higher pH values of the lower horizons are taken into account, is the importance of chelating acids such as low-molecular aliphatic (eg. citric, oxalic) and aromatic acids (eg. vanillic, hydroxy benzoic) as well as fulvic acids. Although no information can be given on the quantity and nature of these various chelating compounds, the soil waters with the relatively high aluminium and iron concentrations all have clear yellow to brown-yellow colours, indicating the importance of the soluble organic compounds. Most of the aluminium and trivalent iron in solution will be present in chelated form. Therefore the activities of the aluminium in the soil waters which have been used later on (fig. 27) will be probably several orders of magnitude lower than its total concentration. Little is known about the chelating acids in soil waters. Schnitzer and co-workers (see Schnitzer & Khan 1972) found that between pH 3-5 fulvic acid (FA) forms molar 1:1 complexes with divalent metal ions (Schnitzer & Hansen 1970). Identical FA/metal molar ratios have been established for Al^{3+} and Fe^{3+} complexes at somewhat lower pH levels (1.7 - 2.4). Apparent stability constants of metal-FA complexes increase with increasing pH. Of all metals investigated, Fe^{3+} forms the most stable complex with FA. The order of apparent stabilities at low pH is: $Fe^{3+} > Al^{3+} > Cu^{2+} > Ni^{2+} > Co^{2+} > Pb^{2+} = Ca^{2+} > Zn^{2+} > Mn^{2+} > Mg^{2+}$ (Schnitzer & Hansen 1970). The $\log K_{app}$ of the divalent metal/FA complexes is about 3 to 4 units lower than those of the Fe^{3+}/FA complex at pH 3.0 and 5.0. The increase in the $\log K_{app}$ values by increasing pH for iron and aluminium/FA complexes is much higher than those for most other important metal ions such as magnesium and calcium. Therefore it can be concluded that at low pH, chelation in soil waters of metal ions such as magnesium, sodium, potassium and to somewhat lower extent calcium, is relatively unimportant with respect to chelation of aluminium and iron. The

soil solution seems to be effective to some extent,
resulting in a slight underestimation of the amount of
K-mica destroyed in the weathering environment, while
somewhat more magnesium and aqueous silica has to be used.
One problem has not been considered. The mass balance of
aqueous silica does not give completely satisfactory
results (table 21): the negative value for $H_4SiO_4^0$
indicates a slightly lower mobilization of aqueous silica
than might be expected from the alteration reactions.
Some speculations can be made on the behaviour of aqueous
silica. One possibility is the precipitation of quartz.
The springwaters are supersaturated with respect to quartz
and the X-ray analyses and norm calculations (see B 1.2)
indicate a (residual) enrichment of this mineral in the
soils. However, quartz certainly does not govern the
aqueous silica activities in the springwater and therefore
it is unlikely that quartz plays an important role in the
silica budget in table 21. The silica budget in table 21
is still better in the case of the chance back reaction
"kaolinite + ferric iron + magnesium → chlorite, rich in
ferrous iron + silica" than in the case of the change back
reactions "kaolinite + magnesium → Mg-chlorite + silica,
and geothite → siderite". Both of the chlorite
compositions have been developed by norm calculations and
represent only approximative compositions, although these
two types of chlorite have been established in the Haarts
catchment. Also the results of table 21 indicate that the
greater part of the magnesium liberated by weathering is
coming from the chlorite rich in ferrous iron. The
relatively fast alteration of this mineral is confirmed
by the chemical analysis of the weathering materials
(see B 1.2). However, no definite conclusion can be
drawn on the precise contribution of this chlorite and that
of the Mg-chlorite to the magnesium levels in the springwaters, until the Mg-content of siderite is known.

Consequently, if all these uncertainities in the mass
balance are taken into account, and the surface soil subsystem with Al-saturated clays and Al-interlayering
(see B 1.2) is excluded, it seems to be justified to conclude that chemical weathering of the low-grade metamorphic
shales in the Haarts catchment can be described in terms
of simple dissolution-precipitation reactions. This overall
weathering system results in the leaching of alkali and
alkaline earth cations, aqueous silica and the formation of
kaolinite, while at periods of low flow rates of the
aqueous solution, partial equilibria between kaolinite and
vermiculite and/or smectite occur. Especially the contribution of albite and chlorite rich in ferrous iron seems to
be relatively high if the molar ratios of sodium and
magnesium in the springwaters are taken into account.

2.2.2 Soil water chemistry
Although the springwaters in the Haarts catchment
clearly represent water which is coming from the soils, the
term soil water is reserved for those waters which were
collected by porous cup soil water samplers and by PVC
drains both installed in the soil profiles. As mentioned

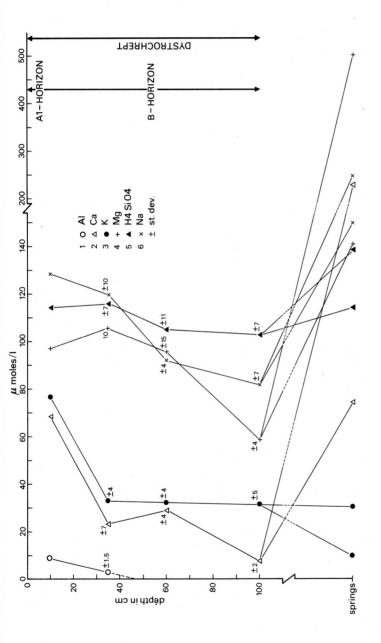

Figure 26　Solutes concentration in soil water samples (porous cups) arranged at various depths, and in springwaters (min. and max. concentrations)

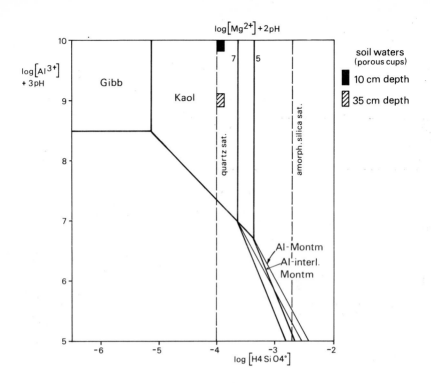

<u>Figure 27</u> Stability fields of gibbsite, kaolinite, Al-
montmorillonite, Al-interlayered montmorillonite
and solution (298.15^{0}K, 1 bar pressure)
(see also table 23)

difference in chelating power for the above mentioned
elements probably explains the much lower decrease of
magnesium and sodium concentrations in the soil waters of
the surface horizons, going towards deeper horizons
(fig. 26). The strong decrease in the magnesium
concentrations for the subsurface horizon in the Dystrochrept can be explained by recycling through plants.
Potassium concentration levels show a strong decrease
going from the Al (10 cm depth) to the cambic horizon
(35 cm depth) and reach a low but almost constant level in
this subsurface horizon. These constant potassium
concentrations cannot be explained by a partial equilibrium
with a K-containing mineral such as illite because the soil
waters are definitely undersaturated with respect to this
mineral. Because at the observed pH levels, chelating
of potassium in the surface horizon plays at most a very
subordinate role, the decrease of potassium concentrations
can only be attributed to a selective uptake by plant roots
and/or adsorption at soil particles. Also the potassium
concentrations in the throughfall waters often show higher
levels than those in the deeper soil waters. So it can be

concluded that (most of the) potassium is continuously recycled in the ecosystem.

Calcium concentration levels show similar trends for the surface horizon soil waters as potassium levels. At the adsorption complex of the A1 horizons also higher calcium levels have been established. This has often been described in Dystrochrepts under forest in the temperate humid climate and is probably due to the relatively high calcium content in fallen leaves and litter material (Duchaufour 1977). Although calcium is somewhat stronger chelated by fulvic acids than magnesium and monovalent metal ions, the higher calcium concentration levels in the A1 horizon and their strong decrease in the subsurface horizon, probably has to be explained by adsorption/desorption processes and recycling by the biomass.

Aqueous silica concentration of porous cup soil waters from various depths shows very small fluctuations (fig. 26). A very slight decrease in these levels has been observed, but due to the small amount of samples in the uppermost horizon no statistically significant differences could be established. These aqueous silica levels seem to be affected by adsorption/desorption mechanisms and/or leaching of silica from plant remnants. The contribution of rock and soil forming minerals to the aqueous silica concentrations in the upper part of the soil seems to be limited.

The soil waters collected with the PVC drains in the A1 horizons show strong fluctuations in the aqueous silica levels (fig. 22a). Probably these levels seem to be governed by the flow rate of the percolating soil water, which is very high in the uppermost horizon. Freely drained soil water will reach the bottom of the A1 horizons within one minute, if this soil horizon is saturated with water.

Although no data are available with respect to the chelating effect in the soil waters of the surface horizons, an attempt will be made to evaluate the rock(soil)-water interactions. No correction has been made for the metal complexing compounds and consequently the data on the aluminium, magnesium and potassium activities used in fig. 27 and 22a are too high. This will be the case especially for the aluminium activities of the coloured soil waters (10 and 35 cm depth). As already discussed, A1-chelates with a molar metal/FA ratio of 1 are mobile, but those with higher molar metal/FA ratios (3-6) have a very low solubility according to Schnitzer & Skinner (1963, 1964). These authors state that organic acids penetrating with the rainwater into the soil gradually take up aluminium and iron in the hydrolized form on their way downwards and become increasingly "saturated" and precipitate at certain depth (Mohr *et al* 1972). Consequently these chelates do not move far and iron and aluminium mobilization is limited to the upper soil horizons. Although no accumulation of free iron and aluminium was observed in the soil profiles at certain depths, the soil water compositions corroborate this statement. At this

Table 23 Equations and constants used in construction
of fig. 27

mineral	equation	source
quartz	$\log(H_4SiO_4^0) = -4.00$	1
amorphous silica	$\log(H_4SiO_4^0) = -2.71$	2
gibbsite	$\log(Al^{3+}) + 3pH = 8.48$	1
kaolinite	$\log(Al^{3+}) + 3pH + \log(H_4SiO_4^0) = 3.33$	1
Al-montmorillonite	$3.35(\log(Al^{3+}) + 3pH) + 0.58(\log(Mg^{2+}) + 2pH) + 7.87\log(H_4SiO_4^0) = -1.25$	3
Al-interlayered montmorillonite	$4.28(\log(Al^{3+}) + 3pH) + 0.58(\log(Mg^{2+}) + 2pH) + \log(H_4SiO_4^0) = 5.26$	3

1   Helgeson (1969)

2   Stumm & Morgan (1970)

3   recalculated by van Breemen (in Brinkman 1979) from
data by Kittrick (1971) assuming equilibrium with
hematite and Al-saturation under Kittrick's experimental
conditions, and assuming the following cation exchange
relationships:

$$exch\text{-}Ca_3 + 2Al^{3+} \rightleftharpoons exch\text{-}Al_2 + 3Ca^{2+} \qquad \log K = 0.78$$

$$exch\text{-}Al + 5Al^{3+} + 15H_2O = exch\text{-}(Al_2(OH)_5)_3 + 15H^+$$
$$\log K = 35$$

The latter constant implies stability of interlayers
with respect to gibbsite due to a degree of solid
solution (not shown)

depth microbiological degradation of the organic acids
probably plays an important role resulting in locally high
Al^{3+} activity (Brinkman 1979). These environmental
conditions probably favour the formation of Al-interlayered
minerals, which have been observed in the B2 and B3 horizons
of the Dystrochrepts in the Haarts catchment (see B 1.2).
However, the stability of these interlayered minerals could
not be affirmed by plotting the $\log(Al^{3+})$ + 3pH and
$\log(H_4SiO_4^0)$ values in a stability diagram (fig. 27).
Although the $\log(Al^{3+})$ + 3pH values are certainly too
high, due to the chelating effects, also the silica
activities seem to be too low in the soil environment to
confirm that an Al-interlayered mineral will be thermo-

dynamically stable phase. However, one has to bear in mind that the stability diagram (fig. 27) has been constructed with a rather preliminary set of thermochemical data (table 23), which was taken from Brinkman (1978).

From the X-ray investigations and chemical analyses of the soil profiles (B 1.2) it could be established that a considerable amount of ferrous iron is trapped in the interlayer positions. This might increase the stability of the interlayered minerals with respect to kaolinite and in the stability diagram (fig. 27), although this process was not taken into account. Therefore no definite conclusion can be presented on the stability of the Al-interlayered mineral in the upper part of the soil environment.

3. MASS TRANSFER CALCULATION MODELS IN WEATHERING PROCESSES

3.1 Introduction

Numerous investigations have been made in recent years of equilibrium models in geochemical systems. However, few attempts have been made to predict the extent to which components in a system are redistributed by geochemical processes. In chemical petrology, the principles involved in making such predictions in equilibrium models are well established (Korzhinskii 1959, 1965; Thompson 1955, 1959; Zen 1963, 1966). Korzhinskii (1963, 1964, 1965) integrated these redistribution principles with solution chemistry and applied them quantitatively to specific irreversible processes in geologic systems. Helgeson (1968), Helgeson et al (1969), Helgeson et al (1970) and Helgeson (1971) applied Korzhinskii's principles to the calculation of mass transfer, reaction paths and chemical implications of irreverisibility in idealized models of weathering, diagenesis, hydrothermal rock alteration and ore deposition.

Natural processes are irreversible processes in an overall sense. That is the thermodynamic systems involved together with their surroundings cannot be restored to their initial states without producing changes in the rest of the universe (Helgeson 1968). However, any geochemical process can be represented by a succession of partial equilibrium states, each reversible with respect to the next, but all irreversible in relation to the initial state of the system. Partial equilibrium represents a state in which a system is in equilibrium with respect to at least one process or reaction but out of equilibrium with respect to others (Barton et al 1963). There is partial equilibrium in any part of a system in which the phases are not all mutually incompatible (Helgeson 1968). In such a system many equilibria will exist together, with a large difference in relaxation time. Therefore classical (equilibrium) thermodynamics can be used for those processes which aim relatively quickly to their equilibrium state, notwithstanding the fact that other processes are not (yet) in their equilibrium state. An irreversible process that involves a series of successive partial equilibrium states may result in a state of local equilibrium for the system; that is, a state in which no mutually incompatible phases are in contact, even though the system as a whole is not in equilibrium (Thompson 1959). Partial equilibrium is probably maintained in most weathering processes. This is supported by the chemistry of ground- and soil waters involved. Changes in composition in a system resulting from a geochemical process are path dependent functions and therefore they cannot be evaluated by considering only the initial and the final states of the system. Knowledge of, or assumptions about, the relative reaction rates of the metastable and stable partial equilibrium states and components have to be known.

Weathering can be considered as a reaction of a mineral assemblage with an aqueous phase (and a gas phase). The reactions are caused by an aqueous solution which is not in (overall) equilibrium with the mineral assemblage. Therefore mineral transformations by congruent and incongruent solution will occur and the resulting redistribution of the components in the system is controlled by the composition of the aqueous solution and the reactant mineral assemblage, and the stable and metastable partial equilibrium states established during the process (Helgeson 1968). A thermodynamic framework is necessary to evaluate the effects of this control mechanism in geologic systems. Such a thermodynamic framework will not be presented in this study, because it has already been well documented (Korzhinsky 1963, 1964, 1965; Helgeson 1968; Fritz 1975).

3.2   The evaluation of the weathering model

In order to evaluate a weathering model several data have to be known and/or some assumptions have to be made. Theremochemical data (Gibbs free energy of formation) of the minerals of the rock and alteration products are needed to be able to establish the stabilities of these minerals as functions of the composition of the aqueous solution. Also data on Gibbs free energy of formation of the aqueous species have to be known. Changes in the composition of an aqueous solution reacting with a rock can be represented quantitatively on an equilibrium activity if the initial composition of the aqueous phase has been specified (ie. the precipitation water in equilibrium with a certain partial CO_2 pressure), as well as the calculation of the mole transfer in the weathering processes involved.

The following assumptions have been made for the weathering model:

1.   An equilibrium exists between the dissolved species in the system. This means that as soon as the solubility product of a solid phase is exceeded, the mineral concerned precipitates. In other words, the partial equilibrium concept has been used.

2. The system is closed with respect to exchanges of
    liquid and solid matter. This means that the newly
    formed solid phases will stay in contact with the
    aqueous solution and will react in the following step
    of the geochemical process.

3. The weathering system is open to the (soil)
    atmosphere. This means that mass transfer calculations
    have to be performed for a limited range of partial
    CO_2 pressures. The highest and lowest partial CO_2
    pressures ($10^{-1 \cdot 65}$ and $10^{-2 \cdot 30}$ bars) obtained from pH
    and alkalinity measurements of spring waters were used
    in the model.

4. Alteration of the mineral assemblage under standard
    conditions (298.15^0K and 1 bar total pressure) is
    considered. This means that the influence of
    temperature fluctuations in the system has been ignored.

5. Inert components like silica as quartz and, at a
    certain stage of the weathering process, aluminium,
    will be conserved in the system. This means that the
    total mass of the inert component supplied by the
    reactant mineral is assumed to go into the product
    mineral rather than into the aqueous phase. The
    chemical potential of the inert component is specified
    by the partial equilibrium states in the system, and
    it changes to whatever extent is necessary to maintain
    these partial equilibrium states as the irreversible
    reaction proceeds. If the inert component is con-
    served in the reaction, the corresponding changes in
    the mass transfer of this component between the product
    mineral and the aqueous solution are neglected. These
    mass transfer changes are often negligible for aluminium
    and silica as quartz and also sometimes for other
    components (ferric iron) in most irreversible reactions
    involving partial equilibrium between silicates and an
    aqueous solution in geochemical processes (Helgeson
    1968).

6. Relative reaction rates have to be known for the
    minerals in the rock. Almost nothing is known about
    these relative reaction rates for most silicates in
    a mineral assemblage. Therefore it was assumed
    firstly that the relative reaction rates of the
    minerals, with the exception of inert minerals like
    quartz and rutile, are proportional to the ratio of
    the mole fractions of these minerals, and secondly
    that this ratio does not change upon alteration.
    This latter assumption cannot be maintained throughout
    the process, but can be accounted for.

7. Weathering in this model can be considered as a stepwise
    reaction of an incremental addition of small amounts
    of rock to 1 kg of water in equilibrium with a
    certain partial CO_2 pressure.

    In the area investigated the bulk composition and
the mineral assemblage of the rock shows a relatively
restricted variation. Rocks could be distinguished on the

basis of their CO_2 content as well as in the variations
in the quartz and to a lesser degree in the muscovite
(sericite) contents.

One rock "type" shows a relatively high CO_2 content
(ca. 3.5 %) indicating that siderite is an important component of this rock. This group of rock consists of quartz,
K-mica (muscovite), albite, siderite and a chlorite
(probably without or with little ferrous iron), hematite,
rutile and traces of apatite and pyrite. The second rock
type that could be established has a very low CO_2 content,
indicating at most a small amount of siderite. It also
has a ferrous iron content similar to the first rock type,
probably indicating that the chlorite in these rocks
contains a considerable amount of ferrous iron which has
often been established in similar rocks of the same metamorphic facies (greenschist facies) (Turner 1968; Muller &
Saxena 1977). Within this rock type shales (phyllites)
and sandy shales (quartzophyllites) can be distinguished
(see B 1.1.2). All of the rock types contain very small
amounts of apatite and traces of pyrite. These constituents
were neglected in the weathering model because their contribution to the weathering processes is negligible according
to assumption (6). That neglecting of pyrite is
acceptable is corroborated by the fluctuations of the sulphate
concentrations in spring waters throughout the year. A
slightly negative relationship ($r = -0.20$) exists between
these sulphate levels and the alkalinity, of which the latter
is an overall measure for the weathering intensity. So it
can be concluded that the sulphate levels of the springwaters
are governed by the sulphate input by precipitation and that
the contribution of the pyrite by weathering to these levels
is at most of minor importance. Representative members of
the two relevant rock types were selected for evaluating
the mass transfer changes of the reactant minerals and
the reaction pathways caused by weathering, described in the
next few paragraph.

3.3   Weathering models of the low-grade metamorphic rocks

Before describing the weathering model, a set of data
has to be built up and with some of these data equilibrium
activity diagrams have to be constructed (table 25) to
show the stability relations among the phases identified
in the rocks, soils and weathering materials and to depict
the reaction paths. The system for both rock "types"
contains 9 components ie. MgO - K_2O - Na_2O - SiO_2 - Al_2O_3 -
FeO - Fe_2O_3 - $H_2O(1)$ - $CO_2(g)$ and various equilibrium
activity diagrams were constructed. To enable drawing of
these diagrams a division was made into two subsystems
viz. an MgO - K_2O - (Na_2O) - SiO_2 - Al_2O_3 - $H_2O(1)$ - $CO_2(g)$
system, which is represented in the figures 28 and 29 and
a second one for the FeO - Fe_2O_3 - $H_2O(1)$ - $CO_2(g)$ system.
The thermodynamic data used for the construction of these
diagrams and the equilibrium reactions used in the mass
transfer calculations are given in tables 24, 25, 26, 27.

The process of weathering of the rock can be regarded
as a combination of hydrolysis reactions, redox reactions
and acid attack by CO_2-charged soil waters. Therefore,

Table 24 Gibbs free energies of formation of minerals and species in the weathering model

Mineral/species	Formula	ΔG^{o}_{f} kcal/gfw	Source
Kaolinite	$Al_2Si_2O_5(OH)_4$	-903.0	Helgeson (1969)
Gibbsite	$Al(OH)_3$	-273.5	"
Illite	$K_{0.6}Mg_{0.25}Al_{2.3}Si_{3.5}O_{10}(OH)_2$	-1301.0	"
Mg-Beidellite	$Mg_{0.167}Al_{2.33}Si_{3.67}O_{10}(OH)_2$	-1275.34	"
Clinochlore	$Mg_5Al_2Si_3O_{10}(OH)_8$	-1974.0	Zen (1972)
Vermiculite	$K_{0.10}Mg_{6.69}Fe^{III}_{0.59}Al_{1.61}Si_{5.98}O_{20}(OH)_4$	-2710.5	Verstraten (1979b)
Muscovite	$KAl_3Si_3O_{10}(OH)_2$	-1330.10	Robie and Waldbaum (1968)
Low Albite	$NaAlSi_3O_8$	-883.99	"
Hematite	Fe_2O_3	-177.7	"
Goethite	$FeOOH$	-117.7	Mohr et al. (1972)
Siderite	$FeCO_3$	-159.34	Robie et al. (1978)
Magnesite	$MgCO_3$	-246.05	"
Amorph.iron	$Fe(OH)_3$	-166.5	Langmuir (1971)
Mg^{2+}		-108.90	Robie and Waldbaum (1968)
K^+		-67.70	"
Na^+		-62.54	"
Al^{3+}		-115.0	Latimer (1952)
H_4SiO_4		-312.7	Robie and Waldbaum (1968)
$CO_2(g)$		-94.26	"
H_2O		-56.69	"
OH^-		-37.59	"
H^+		0	"
Fe^{2+}		-18.85	Langmuir (1971)
$FeOH^+$		-64.22	"
Fe^{3+}		- 1.1	"

the weathering process cannot be represented by a simple hydrolysis model (Helgeson et al 1969). The range of partial CO_2 pressures, obtained from pH and alkalinity measurements lies between $10^{-1.65}$ and $10^{-2.30}$ bars. This means that, if the influence of the acidity of the precipitation is ignored, which seems to be acceptable (see B 4.2), weathering takes place by an aqueous solution

Table 25 Equilibrium reactions for constructing the equilibrium activity diagrams[*]

1. Kaolinite/Gibbsite	$\lg K = 2\lg[H_4SiO_4^0] = 10.30$
2. Chlorite/Gibbsite	$\lg K = 5\{\lg[Mg^{2+}]+2pH\}+3\lg[H_4SiO_4^0]= 40.76$
3. Chlorite/Kaolinite	$\lg K = 5\{\lg[Mg^{2+}]+2pH\}+\lg[H_4SiO_4^0]=51.06$
4. Vermiculite/Kaolinite	$\lg K = 13.38\{\lg[Mg^{2+}]+2pH\}+0.20\{\lg[K^+]+pH\}+8.74\lg[H_4SiO_4^0]=109.92$[**]
5. Verm./Mg-Beidellite	$\begin{cases}\lg K=15.3188\{\lg[Mg^{2+}]+2pH\}+0.233\{\lg[K^+]+pH\}+8.024\lg[H_4SiO_4^0]=\\132.92\text{[**]}\end{cases}$
6. Vermiculite/Chlorite	$\lg K = 5.33\{\lg[Mg^{2+}]+2pH\}+0.20\{\lg[K^+]+pH\}+7.13\lg[H_4SiO_4^0]=27.71$[**]
7. Mg-Beidellite/Kaol.	$\lg K = 0.33\{\lg[Mg^{2+}]+2pH\}+2.68\lg[H_4SiO_4^0]=-6.35$
8. Illite/Kaolinite	$\lg K = 0.5\{\lg[Mg^{2+}]+2pH\}+1.2\{\lg[K^+]+pH\}+2.4\lg[H_4SiO_4^0]=2.88$
9. Illite/Mg-Beidellite	$\lg K = 0.198\{\lg[Mg^{2+}]+2pH\}+1.398\{\lg[K^+]+pH\}-0.286\lg[H_4SiO_4^0]=10.255$
10. Vermiculite/Illite	$\begin{cases}\lg K=14.9845\{\lg[Mg^{2+}]+2pH\}-0.736\{\lg[K^+]-pH\}+8.119\lg[H_4SiO_4^0]=\\124.10\text{[**]}\end{cases}$
11. Illite/Gibbsite	$\lg K = 0.25\{\lg[Mg^{2+}]+2pH\}+0.6\{\lg[K^+]+pH\}+3.5\lg[H_4SiO_4^0]=-10.41$
12. Chlorite/Illite	$\lg K = 11\{\lg[Mg^{2+}]+2pH\}-1.2\{\lg[K^+]+pH\}-0.1\lg[H_4SiO_4^0]=114.57$
13. Muscovite/Gibbsite	$\lg K = \lg[K^+]+pH+3\lg[H_4SiO_4^0]=-10.27$
14. Muscovite/Kaolinite	$\lg K = \lg[K^+]+pH=5.18$
15. Vermiculite/Muscovite	$\lg K = 20.07\{\lg[Mg^{2+}]+2pH\}-1.31\{\lg[K^+]+pH\}+13.11\lg[H_4SiO_4^0]=156.55$[**]
16. Mg-Beidellite/Muscovite	$\lg K = 0.5\{\lg[Mg^{2+}]+2pH\}-2.33\{\lg[K^+]+pH\}+4.02\lg[H_4SiO_4^0]=-21.20$
17. Albite/Kaolinite	$\lg K = 2\{\lg[Na^+]+pH\}+4\lg[H_4SiO_4^0]=0.51$
18. Albite/Gibbsite	$\lg K = \lg[Na^+]+pH+3\lg[H_4SiO_4^0]=-4.90$
19. Quartz saturation	$\lg K = \lg[H_4SiO_4^0]=-4.00$
20. Amorph. silica sat.	$\lg K = \lg[H_4SiO_4^0]=-2.71$
21. Magnesite saturation	$\lg K = \lg[Mg^{2+}]+2pH+\log P_{CO_2}=10.12$
22. Water/Oxygen	$pe = 20.76 - pH$
23. Water/Hydrogen	$pe = -pH$
24. Siderite/Goethite	$pe = 2.98 - pH + \lg P_{CO_2}$
25. Goethite/Fe^{2+}-ions	$pe = 10.64 - 3pH - \lg[Fe^{2+}]$
26. Goethite/$FeOH^+$-ions	$pe = 2.35 - 2pH - \lg[FeOH^+]$
27. Fe^{2+}/$FeOH^+$-ions dom.	$\lg K = \lg[FeOH^+] - \lg[Fe^{2+}] - pH = -8.30$
28. Siderite saturation	$\lg K = \lg[Fe^{2+}] + 2pH + \lg P_{CO_2} = 7.67$
29. Siderite saturation	$\lg K = \lg[FeOH^+] + pH + \lg P_{CO_2} = -0.63$

[*] Brackets denote activities;

[**] $[Fe^{3+}]$ governed by amorphous ferric iron solubility

Table 26 Equilibrium constants of various aqueous species

Aqueous species	Reaction	Log $K_{298.15^\circ K}$, 1 bar pressure
H_2O	$H_2O \rightleftharpoons H^+ + OH^-$	-14.00
$H_2CO_3^*$	$H_2CO_3^* \rightleftharpoons CO_2(g) + H_2O$	1.47
$H_2CO_3^*$	$H_2CO_3^* \rightleftharpoons H^+ + HCO_3^-$	-6.37
HCO_3^-	$HCO_3^- \rightleftharpoons H^+ + CO_3^{2-}$	-10.33
$AlOH^{2+}$	$AlOH^{2+} + H^+ \rightleftharpoons Al^{3+} + H_2O$	4.98
$Al(OH)_2^+$	$Al(OH)_2^+ + 2H^+ \rightleftharpoons Al^{3+} + 2H_2O$	9.74
$Al(OH)_4^-$	$Al(OH)_4^- + 4H^+ \rightleftharpoons Al^{3+} + 4H_2O$	22.04
$H_4SiO_4^\circ$	$H_4SiO_4^\circ \rightleftharpoons H_3SiO_4^- + H^+$	-9.93
$FeOH^{2+}$	$FeOH^{2+} + H^+ \rightleftharpoons Fe^{3+} + H_2O$	2.17
$Fe(OH)_2^+$	$Fe(OH)_2^+ + 2H^+ \rightleftharpoons Fe^{3+} + 2H_2O$	7.17
$Fe(OH)_3^\circ$	$Fe(OH)_3^\circ + 3H^+ \rightleftharpoons Fe^{3+} + 3H_2O$	13.60
$Fe(OH)_4^-$	$Fe(OH)_4^- + 4H^+ \rightleftharpoons Fe^{3+} + 4H_2O$	21.89
$FeOH^+$	$FeOH^+ + H^+ \rightleftharpoons Fe^{2+} + H_2O$	8.30
$Fe(OH)_2^\circ$	$Fe(OH)_2^\circ + 2H^+ \rightleftharpoons Fe^{2+} + 2H_2O$	17.60
$Fe(OH)_3^-$	$Fe(OH)_3^- + 3H^+ \rightleftharpoons Fe^{2+} + 3H_2O$	31.98
$Fe(OH)_4^{2-}$	$Fe(OH)_4^{2-} + 4H^+ \rightleftharpoons Fe^{2+} + 4H_2O$	45.17
$MgOH^+$	$MgOH^+ + H^+ \rightleftharpoons Mg^{2+} + H_2O$	11.40
$MgHCO_3^+$	$MgHCO_3^+ \rightleftharpoons Mg^{2+} + HCO_3^-$	-0.90
$MgCO_3^\circ$	$MgCO_3^\circ \rightleftharpoons Mg^{2+} + CO_3^{2-}$	-3.40
$NaHCO_3^\circ$	$NaHCO_3^\circ \rightleftharpoons Na^+ + HCO_3^-$	0.25
$NaCO_3^-$	$NaCO_3^- \rightleftharpoons Na^+ + CO_3^{2-}$	-1.27
$Na_2CO_3^\circ$	$Na_2CO_3^\circ \rightleftharpoons 2Na^+ + CO_3^{2-}$	0.68
H_2O	$2H_2O \rightleftharpoons O_2(g) + 4H^+ + 4e^-$	-83.12
Fe^{2+}	$Fe^{2+} \rightleftharpoons Fe^{3+} + e$	13.01
Fe^{2+}	$Fe^{2+} + \frac{1}{4}O_2(g) + H^+ \rightleftharpoons Fe^{3+} + \frac{1}{2}H_2O$	7.77

$H_2CO_3^* = H_2CO_3^0 + CO_2^0$

Table 27 Equilibrium constants of the various minerals

Mineral	Reaction	Log $K_{298.15°K}$, 1 bar pressure
Gibbsite	$Al(OH)_3 + 3H^+ \rightleftharpoons Al^{3+} + 3H_2O$	8.48
Kaolinite	$Al_2Si_2O_5(OH)_4 + 6H^+ \rightleftharpoons 2Al^{3+} + 2H_4SiO_4 + H_2O$	6.66
Clinochlore	$Mg_5Al_2Si_3O_{10}(OH)_4 + 16H^+ \rightleftharpoons 5Mg^{2+} + 2Al^{3+} + 3H_4SiO_4 + 6H_2O$	57.73
Mg-Beidellite	$6Mg_{.167}Al_{2.33}Si_{3.67}O_{10}(OH)_2 + 44H^+ + 16H_2O \rightleftharpoons Mg^{2+} + 14Al^{3+} + 22H_4SiO_4$	28.75
Illite	$K_{0.6}Mg_{0.25}Al_{2.3}Si_{3.5}O_{10}(OH)_2 + 8H^+ + 2H_2O \rightleftharpoons 0.6K^+ + 0.25Mg^{2+} + 2.3Al^{3+} + 3.5H_4SiO_4$	9.10
Vermiculite	$K_{0.1}Mg_{6.69}Fe^{III}_{0.59}Al_{1.61}Si_{5.98}O_{20}(OH)_4 + 20.08H^+ \rightleftharpoons 6.69Mg^{2+} + 0.1K^+ + 0.59Fe^{3+} + 1.61Al^{3+}$ $+ 5.98H_4SiO_4 + 0.08H_2O$	62.35
Muscovite	$KAl_3Si_3O_{10}(OH)_2 + 10H^+ \rightleftharpoons K^+ + 3Al^{3+} + 3H_4SiO_4$	15.18
Low Albite	$NaAlSi_3O_8 + 4H^+ + 4H_2O \rightleftharpoons Na^+ + Al^{3+} + 3H_4SiO_4$	3.59
Quartz	$SiO_2 + 2H_2O \rightleftharpoons H_4SiO_4$	-4.00
Amorph.Silica	$SiO_2 + 2H_2O \rightleftharpoons H_4SiO_4$	-2.71
Siderite	$FeCO_3 + H^+ \rightleftharpoons Fe^{2+} + HCO_3^-$	-0.17
Magnesite	$MgCO_3 + H^+ \rightleftharpoons Mg^{2+} + HCO_3^-$	2.28
Brucite	$Mg(OH)_2 + 2H^+ \rightleftharpoons Mg^{2+} + 2H_2O$	16.91
Goethite	$FeOOH + 3H^+ \rightleftharpoons Fe^{3+} + 2H_2O$	-2.36
Hematite	$Fe_2O_3 + 6H^+ \rightleftharpoons 2Fe^{3+} + 3H_2O$	-3.98

with a pH varying between 4.75 and 5.07 and a bicarbonate content ranging between $10^{-4 \cdot 75}$ and $10^{-5 \cdot 07}$ moles/kg of water.

3.3.1 Rocks with a low CO_2 content

The ratio of the mole fractions of the albite, muscovite, chlorite, hematite, siderite and quartz and rutile is 15.5:35:10:5.6:2.25:309:5.3. If the inert components like quartz and rutile are ignored the weathering model can be built up with the (more) reactive part of the mineral assemblage.

The first stage of alteration is always the congruent dissolution of the rock, if the exchange of H^+ for the metal cations at the surface of the reacting minerals is disregarded (Correns & Von Engelhardt 1938). The general equation, neglecting the carbonate ion and -ionpairs, the H_3SiO_4- and polymerized species of aqueous silica, due to the rather low initial pH (see B 3.3), is in this case as follows:

$$(15.5NaAlSi_3O_8 + 35KAl_3Si_3O_{10}(OH)_2 +$$

$$10Mg_{2.4}Fe^{II}_{1.9}Al_{3.6}Si_{2.15}O_{10}(OH)_8 + 2.25FeCO_3 + 5.6Fe_2O_3)(s) +$$

$$593.05H_2O + (641.85 - c - 2d - e - 2f - 4g - h - 2i - 3j)CO_2(g) \rightarrow$$

$$(15.5 - a)Na^+ + aNaHCO_3^0 + 35K^+ + (24 - b)Mg^{2+} + bMgHCO_3^+ + (21.25 - c - d)Fe^{2+}$$

$$+ cFeOH^+ + dFe(OH)_2^0 + (156.5 - e - f - g)Al^{3+} + eAlOH^{2+} + fAl(OH)_2^+ +$$

$$gAl(oh)_4^- + (11.2 - h - i - j)Fe^{3+} + hFeOH^{2+} iFe(OH)_2^+ + jFe(OH)_3^0 +$$

$$173H_4SiO_4^0 + (6.44.1 - a - b - c - 2d - e - 2f - 4g - h - 2i - 3j)HCO_3^- \qquad (1)$$

The mass transfer and the reaction pathway are calculated as follows (Dirven et al 1976). The pH is increased stepwise. After each step the coefficients a, b, c, etc. of equation (1), using the appropriate thermodynamic relationships of table 26, are calculated and the various concentrations of the dissolved species can be estimated. So it can be concluded from equation (1) that a number of partial equilibrium states, ie. for the ferrous and ferric iron, the aluminium species and the H^+ and OH^- can be assumed to hold in solution, and these equilibria determine the reaction coefficients in the rock dissolving reaction. Checking after each step whether or not the solubility product of one of the relevant solid phases is surpassed if the irreversible reaction proceeds, the weathering path can be constructed. If the solubility product of a certain mineral is exceeded, the equation (1) is modified to accommodate production of this phase. In this case the first solid product that precipitates is geothite (αFeOOH), as already at a total Fe(III) concentration of $10^{-13 \cdot 8}$ moles/kg of water (pH = 4.75) to $10^{-14 \cdot 3}$ moles/kg of water (pH = 5.07), the solubility of geothite is surpassed. This means that the bicarbonate increase is completely negligible in relation to the bicarbonate content in the initial aqueous solution and no mass transfer had to be calculated for this first reaction. So the general equation, also ignoring some still quantitatively unimportant ferrous iron species for the rock dissolution becomes:

$(15.5NaAlSi_3O_8+35KAl_3Si_3O_{10}(OH)_2+$

$10Mg_{2.4}Fe^{II}_{1.9}Al_{3.6}Si_{2.15}O_{10}(OH)_8+2.25FeCO_3+5.6Fe_2O_3)(s)+$

$581.25H_2O+(608.25-c-d-2e-4f)CO_2(g)\rightarrow$

$(15.5-a)Na^++aNaHCO_3^0+35K^++(24-b)Mg^{2+}+bMgHCO_3^++(21.25-c)Fe^{2+}+$

$cFeOH^++(156.5-d-e-f)Al^{3+}+dAlOH^{2+}+eAl(OH)_2^++fAl(OH)_4^-+$

$173H_4SiO_4^0+(610.5-a-b-c-d-2e-4f)HCO_3^-+11.2FeOOH(s)$ (2)

As reaction (2) proceeds the aqueous solution becomes more concentrated and for both partial CO pressures the solubility product of gibbsite will be surpassed first. The point at which the solution becomes saturated with respect to gibbsite was calculated with the Davis equation (Davis 1962) to convert molalities into activities. So a perfect equilibrium between the dissolved aluminium species and gibbsite is assumed to exist. This point is indicated in the figure 28 as P and A for the P_{CO_2} = $10^{-2.30}$ and $10^{-1.65}$ bars and in table 28 and 29 the compositions of the aqueous solutions and the mass transfer are given.

From the moment that gibbsite precipitates the weathering of the rock for both partial CO pressures proceeds according to:

$(15.5NaAlSi_3O_8+35KAl_3Si_3O_{10}(OH)_2+$

$10Mg_{2.4}Fe^{II}_{1.9}Al_3.6Si_{2.15}O_{10}(OH)_8+2.25FeCO_3+5.6Fe_2O_3)(s)+$

$581.85H_2O+(138.75-c)CO_2(g)\rightarrow$

$(15.5-a)Na^++aNaHCO_3^0+35K^++(24-b)Mg^{2+}+bMgHCO_3^++(21.25-c)Fe^{2+}+$

$cFeOH^++173H_4SiO_4^0+(141-a-b-c)HCO_3^-+156.5Al(OH)_3(s)+11.2FeOOH(s)$
     (3)

In most weathering systems the activities of the aluminium species in the aqueous solution are so low that even large relative changes in their activities do not considerably affect the mass transfer among the solids in the system. Therefore such a process can be considered as one in which aluminium is essentially conserved among the solid phases (Helgeson 1968). This aluminium conservation principle was used for equation (3). Nevertheless, actually a small amount of aluminium is transferred to the aqueous solution to maintain partial equilibrium states represented by the various hydrolitic equilibria of the dissolved aluminium species (table 26) as well as that between gibbsite and the aqueous solution.

If the rock weathering proceeds according to equation (3) the changes in the composition of the aqueous solution are given by the curves PQ and AB for the two partial CO_2 pressures and the increase in the H_4SiO_4 concentrations result in a partial equilibrium between kaolinite and the solution being established and gibbsite becoming unstable. The composition of the aqueous solutions and the mass transfer at these points are given in tables 28 and 29. If all partial equilibrium states are maintained and the rock is in excess, which is of course normally the case, the gibbsite produced in the interval PQ and AB will react with the rock and kaolinite is formed. The only way to fulfil the requirements is a constant acitivity of the $H_4SiO_4^0$ and therefore the reaction, with aluminium conserved, an be written as follows:

$$(15.5NaAlSi_3O_8+35KAl_3Si_3O_{10}(OH)_2+$$

$$10Mg_{2.4}Fe^{II}_{1.9}Al_{3.6}Si_{2.15}O_{10}(OH)_8+2.25FeCO_3+5.6Fe_2O_3)(s)+$$

$$16.5Al(OH)_3(s)+149.35H_2O+(138.75-c)CO_2(g)\rightarrow$$

$$(15.5-a)Na^++aNaHCO_3^0+35K^++(24-b)Mg^{2+}+bMgHCO_3^++(21.25-c)Fe^{2+}+$$

$$cFeOH^++(141-a-b-c)HCO_3^-+11.2FeOOH(s)+86.5Al_2Si_2O_5(OH)_4(s)$$

$$(4)$$

Reaction (4) continues until all of the gibbsite produced in the intervals PQ and AB has been consumed. This will be the case in points R and C for the two partial CO_2 pressures (fig. 28).

The composition of the aqueous solution and the mass transfers in these points are presented in tables 28 and 29. From now on kaolinite begins to form as a weathering product of the incongruent dissolution of the rock by the aqueous solution.

With the aluminium again conserved, this reaction can be written as:

$$(15.5NaAlSi_3O_8+35KAl_3Si_3O_{10}(OH)_2+$$

$$10Mg_{2.4}Fe^{II}_{1.9}Al_{3.6}Si_{2.15}O_{10}(OH)_8+2.25FeCO_3+5.6Fe_2O_3)(s)+$$

$$190.6H_2O+(138.75-c)CO_2(g)\rightarrow$$

$$(15.5-a)Na^++NaHCO_3^0+35K^++(24-b)Mg^{2+}+bMgHCO_3^++(21.25-c)Fe^{2+}+$$

$$cFeOH^++16.5H_4SiO_4^0+(141-a-b-c)HCO_3^-+78.25Al_2Si_2O_5(OH)_4(s)+$$

$$11.2FeOOH(s)$$

$$(5)$$

The coefficients in equation (5) can be calculated from the appropriate thermodynamic relationships (table 26) and the activity coefficients. As a result of equation (5) the composition of the aqueous solution changes respectively from C to D and from R to S (fig. 28 and 29), in the first case for P_{CO_2} $10^{-1.65}$ and in the second case for $10^{-2.30}$ bars.

A part of the silica which is liberated from the rock is withdrawn from the aqueous solution by the precipitation of kaolinite and it is clear that the H_4SiO_4 concentrations increase (fig. 28 and 29) and the aqueous solution becomes

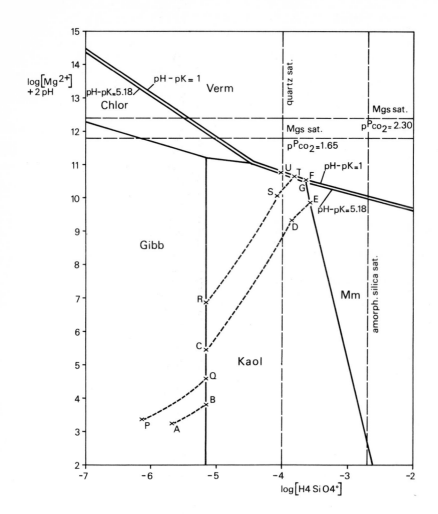

Figure 28   Stability fields and reaction pathways at
             P_{CO_2} = $10^{-2 \cdot 30}$ and $10^{-1 \cdot 65}$ bars for the rocks
             with a low CO_2 content in the system
             $MgO-K_2O-Al_2O_3-Fe_2O_3-SiO_2-H_2O-CO_2(g)$ (298.15^{0}K,
             1 bar total pressure). Amorphous ferric iron
             governs (Fe^{3+}).
             Gibb = gibbsite; Kaol = kaolinite;
             Mm = Mg-beidellite; Verm = vermiculite;
             Chlor = chlinochlore; Mgs = magnesite
             The reaction pathways are PQRSTU for P_{CO_2} =
             $10^{-2 \cdot 30}$bar; ABCDEFG for P_{CO_2} = $10^{-1 \cdot 65}$bar.

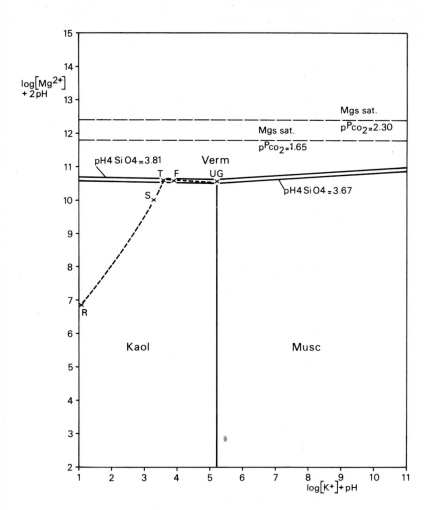

<u>Figure 29</u>   Stability fields and part of the reaction
pathways at P_{CO_2} = $10^{-2\cdot30}$ and $10^{-1\cdot65}$ bars for
the rocks with a low CO_2 content in the system
$MgO-K_2O-Al_2O_3-Fe_2O_3-SiO_2-H_2O-CO_2$ (g) (298.15°K,
1 bar total pressure). Amorph. ferric iron
governs (Fe^{3+}).
Gibb = gibbsite; Kaol = kaolinite; Mm =
Mg=beidellite; Verm = vermiculite; Chlor =
clinochlore; Mgs = magnesite.
The reaction pathways are PSTU for P_{CO_2} =

$10^{-2\cdot30}$bar; FG for P_{CO_2} = $10^{-1\cdot65}$ bar.

supersaturated with regard to quartz. Although new formation of quartz from supersaturated aqueous solutions at room temperatures has been established (Mackenzie & Gees 1971), the precipitation of quartz under standard conditions is a very slow process and it does not govern the aqueous silica activities in aqueous solutions under natural conditions. Therefore it was assumed that quartz is not formed in the weathering model, even though it is a thermodynamically stable phase.

At D and S the solubility product of siderite, $FeCO_3$ is surpassed, provided that, amongst other things, a partial equilibrium state between the aqueous solution and siderite is assumed. The composition of the aqueous solution and the mass transfers at this point are given in tables 28 and 29. From this moment on the pe is determined by the redox couple $FeCO_3(s)/\alpha FeOOH(s)$ and the pH. Upon further weathering of the rock, siderite plays only a sub-ordinate role. Consequently the assumption concerning the ratio of the reaction rates (assumption 6) is no longer operative and it is assumed that the reactions from D and S now follow the equation:

$$(15.5NaAlSi_3O_8+35KAl_3Si_3O_{10}(OH)_2+10Mg_{2.4}Fe_{1.9}^{II}Al_{3.6}Si_{2.15}O$$

$$(OH)_8$$

$$+5.6Fe_2O_3+169.35H_2O+117.5CO_2(g)$$

$$(15.5-a)Na^++aNaHCO_3^{?}+35K^++(24-b)Mg^{2+}+bMgHCO_3^++16.5H_4SiO_4^0+$$

$$(98.5-a-b)+78.25Al_2Si_2O_2(OH)_4(s)+11.2FeOOH(s)+19FeCO_3(s)$$

$$(6)$$

These parts of the weathering paths are represented in the figure 28, and by DT and SE for $P_{CO_2} = 10^{-1.65}$ and $10^{-2.30}$ bars, respectively, and the composition of the aqueous solution and the mass transfers are given in table 28 and 29. From now on a distinction has to be made for the weathering paths of the two partial CO_2 pressures.

$P_{CO_2} = 10^{-1.65}$bars
_____

Under these circumstances at E, Mg-beidellite will be formed because its solubility product is threatened to be surpassed and a partial equilibrium state between the aqueous solution, kaolinite and Mg-bedeillite exists. Due to the fact that the parent rock is rather rich in aluminium, the ratio of Si/Al/Mg in this rock permits a simultaneous formation of kaolinite and Mg-beidellite (and not a consumption of kaolinite) upon further rock weathering if the partial equilibrium concept is accepted. So the weathering reactions becomes (Al-conservation):

$p(15.5NaAlSi_3O_8+35KAl_3Si_3O_{10}(OH)_2+$

$10Mg_{2.4}Fe^{II}_{1.9}Al_{3.6}Si_{2.15}O_{10}(OH)_8+5.6Fe_2O_3)(s)+qH_4SiO_4^0+$

$xH_2O+yCO_2(g)\rightarrow$

$(15.5-a)p\ Na^++ap\ NaHCO_3^0+35p\ K^++r\ Mg^{2+}+s\ MgHCO_3^++t\ HCO_3^-+$

$u\ Al_2Si_2O_5(OH)_4(s)+v\ Mg_{0.167}Al_{2.33}Si_{3.67}O_{19}(OH)_2(s)+$

$11.2p\ FeOOH(s)+19p\ FeCO_3(s)$ ⠀⠀⠀⠀⠀⠀⠀⠀⠀⠀7a)

where $v = (156.5p-2u)/2.33 = 67.1674p - 0.8584u$;

⠀⠀⠀$q = 3.67v + 2u - 173p = 73.5043p - 1.1502u$;

⠀⠀$r+s = 24p - 0.167v = 12.7830p - 0.1433v$;

⠀⠀⠀⠀$t = (15.5-a)p + 35p + 2r + s$;

⠀⠀⠀⠀$x = 1.442u - 114.2412p + 0.5ap + 0.5t + 0.5s$, and

⠀⠀⠀⠀$y = 19p + ap + s + t$

⠀⠀⠀⠀The coefficient of the sodium complex, and the ratio of the Mg^{2+} and the $MgHCO_3$ concentrations to each other can be calculated from the appropriate equilibrium reactions (table 26) and the activity coefficients. Rock weathering continues along the kaolinite/Mg-beidellite boundary until the "triple point" F has been reached with the formation of kaolinite and Mg-beidellite and consumption of aqueous silica. The accompanying changes in the aqueous solution and in the mass transfers according to equation (7a) are presented in table 29. At the point F the aqueous solution becomes saturated with respect to vermiculite and upon further weathering of the rock this solid phase is formed. At this point a partial equilibrium is maintained between the aqueous solution, vermiculite, Mg-beidellite and kaolinite. According to the Gibbs phase rule F = C - P + 2, where F is the number of degrees of freedom, C is the number of components (= 9) and P is the number of phases (= 7), the number of independent variables is 4, which reduces to 2 at constant temperature and pressure (assumption 4). Therefore it might be concluded that the composition of the aqueous solution will change upon further rock weathering.

⠀⠀⠀⠀To maintain the state of partial equilibria between the three solid phases and the aqueous solution, magnesium and Mg-beidellite will be consumed and the potassium concentration increased considerably. To satisfy these partial equilibrium requirements, the rock weathering process continues with the reaction (Al-conservation):

$p(15.5NaAlSi_3O_8+35KAl_3Si_3O_{10}(OH)_2+$

$10Mg_{2.4}Fe^{II}_{1.9}Al_{3.6}Si_{2.15}O_{10}(OH)_8+5.6Fe_2O_3)(s)+$

$q\ Mg_{0.167}Al_{2.33}Si_{3.67}O_{10}(OH)_2(s)+a\ Mg^{2+}+b\ MgHCO_3^++$

$(dp+e+19p+b)CO_2(g)+x\ H_2O\rightarrow$

$(15.5-d)p$ Na^++dp $NaHCO_3^0$+c K^++e HCO_3^-+f $H_4SiO_4^0$+

r $Al_2Si_2O_5(OH)_4(s)$+s $K_{0.10}Mg_{6.69}Fe_{0.59}^{III}Al_{1.61}Si_{5.98}O_{20}(OH)_4(s)$+

19p $FeCO_3(s)$+t $FeOOH(s)$ $\hspace{3cm}$ (8a)

where $\quad s = (16.5p+1.34q-c)/4.37;$

$\qquad\quad t = 11.2p-0.59s;$

$\qquad\quad r = (173p+3.67q-5.98s-f)/2;$

$\qquad a+b = 6.69s-24p-0.167q;$

$\qquad\quad c = 35p-0.1s;$

$\qquad\quad f = 173p+3.67q-2r-5.98s;$

$\qquad\quad e = (15.5-d)p+c-2a-b,$ and

$\qquad\quad x = (dp+e+4f+4r+4s+t-150p-2q-b)/2$

The coefficient of the sodium complex and the ratio of the Mg^{2+} and $MgHCO_3^+$ concentrations were calculated from the appropriate equilibrium reactions (table 26) and the activity coefficients. Reaction (8a) proceeds until the solubility product of muscovite is reached (point G). At this point a partial equilibrium between the aqueous solution, kaolinite, vermiculite, smectite and muscovite exists. The Gibbs phase rule predicts that under these conditions F = 1 so that the composition of the aqueous solution still changes upon further rock weathering.

Most of the Mg-beidellite formed in the interval E F will be consumed during this stage. The changes in the composition of the aqueous solution and the mass transfer caused by reaction (8a) are given in table 29. It appears that the shifting in the location of the triple point F mainly due to changes in the potassium activities in the aqueous solution, is too small to be indicated in figure 28.

From now on rock weathering continues without the muscovite of the parent rock being attacked, and theoretically it will be formed mainly because the pH still rises. The formation of muscovite in nature under earth surface conditions is very unlikely and because the $lg(K^+)$ + pH values of the soil and spring water are much lower than the theoretical values predicted by the weathering model at point G, the weathering model was not evaluated for the subsequent stage (see B 3.3.3)

$P_{CO_2} = 10^{-2.30}bars$

Under these circumstances at T, the aqueous solution becomes saturated with respect to vermiculite and a partial equilibrium between the aqueous solution, kaolinite and vermiculite exists. Here also due to the Si/Al/Mg ratio in the parent rock, a simultaneous formation of vermiculite and kaolinite occurs upon further rock weathering according to (Al-conservation):

$$p(15.5\,NaAlSi_3O_8 + 35\,KAl_3Si_3O_{10}(OH)_2 + 10\,Mg_{2.4}Fe^{II}_{1.9}Al_{3.6}Si_{2.15}O(OH)$$

$$+ 5.6\,Fe_2O_3) + a\,Mg^{2+} + b\,MgHCO_3^+ + c\,H_4SiO_4^0 + (dp+f+19p-b)CO_2(g) + H_2O$$

$$p(15.5-d)Na^+ + dp\ NaHCO_3 + e\ K^+ + f\ HCO_3^- + q\ Al_2Si_2O_5(OH)_4(s)$$

$$+ r\ K_{0.10}Mg_{6.69}Fe^{III}_{0.59}Al_{1.61}Si_{5.98}O_{20}(OH)_4(s) + s\ FeOOH(s)$$

$$+ 19p\ FeCO_3(s)$$

$$(7b)$$

where $a+b = 6.69r - 24p$;

$\quad c = 5.98r + 2q - 173p$;

$\quad r = (156.5p - 2q)/1.61$;

$\quad s = 11.2p - 0.59r$;

$\quad e = 35p - 0.1r$;

$\quad f = (15.5-d)p + e + 2a - b$; and

$\quad x = (dp + f + 4q + 4r + s - 150p - b - 4c)/2$

The coefficients of the ion pairs are calculated as indicated above. The reaction (7b) consumes a relatively large amount of magnesium from the aqueous solution and the potassium and sodium concentrations rise significantly. Therefore the "triple point" chlorite/vermiculite/kaolinite will not be reached, but the solubility product of muscovite threatens to be surpassed and a partial equilibrium between muscovite, vermiculite, kaolinite and the aqueous solution will be adjusted as point U. The changes in the composition of the aqueous solution and the mass transfer by reaction (7b) are given in table 28. Upon further rock weathering muscovite from the parent rock will be preserved and theoretically it will also be formed mainly due to changes in the pH. Here again a further evaluation of the weathering model is not given because from now on it gives unrealistic results (see also B 3.3.3).

3.3.2 Rocks with a relatively high CO_2 content

The ratio of the mole fractions of albite, muscovite, chlorite, hematite, siderite and rutile is 11:14:3.2:18: 282:0.25. For this rock also inert components like quartz and rutile are ignored in the weathering model. The model was modified slightly with respect to the first model. It is assumed that the redox potential is governed by a fixed partial oxygen pressure ($P_{O_2} = 10^{-45}$ bars). The values of the partial CO_2 and O_2 pressures and the pH of the initial conditions are determined by the rate of mineralization of the organic material and the exchange of the soil atmosphere with the atmosphere. A P_{O_2} of 10^{-45} bars indicates slightly anaerobic conditions (pH = 4.75/ p =4.76; pH = 8.5/pe = 1.01). Under these conditions ferrous and ferric iron concentrations in the aqueous solution are very low if geothite governs the ferric iron concentrations, so it can be concluded that most of the iron from the parent rock is fixed in the weathering. material as geothite. Because the same principles operate as in the first model, the second model will be only briefly described.

The first stage of weathering is the congruent dissolution of the rock. The general equation, neglecting the presence of carbonate ions and carbonate ion pairs and various dissolved aqueous silica species, is as follows:

$$(11NaAlSi_3O_8+15KAl_3Si_3O_{10}(OH)_2+3Mg_{4.1}Al_{3.8}Si_{2.1}O_{10}(OH)_8+$$

$$18FeCO_3+3.2Fe_2O_3)(s)+285.6H_2O+$$

$$(280-c-2d-e-2f-4g-h-2i-3j)CO_2(g)\rightarrow$$

$$(11-a)Na^{+}+aNaHCO_3^0+14K^{+}+(12.3-b)Mg^{2+}+bMgHCO_3^{+}+(18-c-d)Fe^{2+}+$$

$$cFeOH^{+}+dFe(OH)_2^0+(64.4-e-f-g)Al^{3+}+eAlOH^{2+}+fAl(OH)_2^{+}+$$

$$gAl(OH)_4^{-}+(6.4-h-i-j)Fe^{3+}+hFeOH^{2+}+iFe(OH)_2^{+}+jFe(OH)_3^0+$$

$$81.3H_4SiO_4+(298-a-b-c-2d-e-2f-4g-h-2i-3j)HCO_3^{-} \qquad (1)$$

The first solid phase that precipitates, without an important quantitative increase of the bicarbonate content, is geothite and from now on iron is also fixed as geothite, because at this stage no other ferric solid phases occur. The general equation becomes (Fe conservation):

$$(11NaAlSi_3O_8+14KAl_3Si_3O_{10}(OH)_2+3Mg_{4.1}Al_{3.8}Si_{2.1}O_{10}(OH)_8+$$

$$18FeCO_3+3.2Fe_2O_3)(s)+270.2H_2O+(224.8-d-2e-4f)CO_2(g)+$$

$$(1.5a+1.5b+1.5d+3e+6f-76.8)O_2(g)\rightarrow$$

$$(11-a)Na^{+}+aNaHCO_3^0+14K^{+}+(12.3-b)Mg^{2+}+bMgHCO_3^{+}+(64.4-d-e-f)Al^{3+}+$$

$$dAlOH^{2+}+eAl(OH)_2^{+}+fAl(OH)_4^{-}+81.3H_4SiO_4^0+(242.8-a-b-d-2e-4f)HCO_3^{-}+$$

$$24.4FeOOH(s) \qquad (2)$$

At points P^1 and A^1 (fig. 30) for the $P_{CO_2} = 10^{-2.30}$ and $P_{CO_2} = 10^{-1.65}$ bars, gibbsite precipitates and the composition of the aqueous solution and the mass transfers at these points are given in tables 30 and 31. From now on the weathering of the rock for both CO_2 pressures proceeds according to (Fe and Al conservation):

$$(11NaAlSi_3O_8+14KAl_3Si_3O_{10}(OH)_2+3Mg_{4.1}Al_{3.8}Si_{2.1}O_{10}(OH)_8+$$

$$18FeCO_3+3.2Fe_2O_3)(s)+270.2H_2O+31.6CO_2(g)+4.5O_2(g)\rightarrow$$

$$(11-a)Na^{+}+aNaHCO_3^0+14K^{+}+(12.3-b)Mg^{2+}+bMgHCO_3^{+}+81.3H_4SiO_4^0+$$

$$(49.6-a-b)HCO_3^{-}+64.4Al(OH)_3(s)+24.4FeOOH(s) \qquad (3)$$

At points Q^1 and B^1 gibbsite becomes unstable and will

130

be consumed at constant H_4SiO_4 activity upon further rock weathering according to (Fe and Al conservation):

$(11NaAlSi_3O_8+14KAl_3Si_3O_{10}(OH)_2+3Mg_{4.1}Al_{3.8}Si_{2.1}O_{10}(OH)_8+$

$18FeCO_3+3.2Fe_2O_3)(s)+16.9Al(OH)_3(s)+66.95H_2O+31.6CO_2(g)+$

$4.5O_2(g)\rightarrow$

$(11-a)Na^++aNaHCO_3^0+14K^++(12.3-b)Mg^{2+}+bMgHCO_3^++(49.6-a-b)HCO_3^-+$

$40.65Al_2Si_2O_5(OH)_4(s)+24.4FeOOH(s)$ \hfill (4)

At points R^1 and C^1 all of the gibbsite formed in the intervals P^1Q^1/A^1B^1 has disappeared (tables 30 and 31) and from now on kaolinite begins to form as a weathering product of incongruent dissolution of the rock by the aqueous solution according to (Fe and Al conservation):

$(11NaAlSi_3O_8+14KAl_3Si_3O_{10}(OH)_2+3Mg_{4.1}Al_{3.8}Si_{2.1}O_{10}(OH)_8+$

$18FeCO_3+3.2Fe_2O_3)(s)+109.2H_2O+31.6CO_2(g)+4.5O_2(g)\rightarrow$

$(11-a)Na^++aNaHCO_3^0+14K^++(1.3-b)Mg^{2+}+bMgHCO_3^++16.9H_4SiO_4^0+$

$(49.6-a-b)HCO_3^-+32.2Al_2Si_2O_5(OH)_4(s)+24.4FeOOH(s)$ \hfill (5)

These points of the weathering paths are represented in fig 30 by R^1S^1 and C^1D^1 for $P_{CO_2} = 10^{-2.30}$ and $10^{-1.65}$ bars respectively and the composition of the aqueous solution and the mass transfer are given in table 30 and 31. At S and D a partial equilibrium between the aqueous solution, kaolinite and Mg-beidellite is maintained. Also for this rock, due to its Si/Al/Mg ratio a simultaneous formation of kaolinite and Mg-beidellite occurs upon further rock weathering according to (Fe and Al conservation):

$p(11NaAlSi_3O_8+14KAl_3Si_3O_{10}(OH)_2+3Mg_{4.1}Al_{3.8}Si_{2.1}O_{10}(OH)_8+$

$18FeCO_3+3.2Fe_2O_3)(s)+q\ H_4SiO_4^0+x\ H_2O+g\ CO_2(g)+4.5p\ O_2(g)\rightarrow$

$(11-a)p\ Na^++aNaHCO_3^0+14K^++r\ Mg^{2+}+s\ MgHCO_3^++t\ HCO_3^-+$

$u\ Al_2Si_2O_5(OH)_4(s)+v\ Mg_{0.167}Al_{2.33}Si_{3.67}O_{10}(OH)_2(s)+$

$24.4p\ FeOOH(s)$ \hfill (6)

where $v = (64.6p-2u)/2.33$;

$q = 3.67v+2u-81.3p$

$r+s = 12.3p-0.167v$;

$t = (11-a)p+14p+2r+s$;

$x = 2u+v+0.5ap+0.5s+0.5t-2q-13.8p$; and

$y = t+ap+s-18p$.

Rock weathering according to reaction (6) continues along the kaolinite/Mg-beidellite boundary (fig.30) until the "triple point" T^1, F^1 has been reached under formation of kaolinite and smectite and consumption of aqueous silica. At this point the aqueous solution becomes saturated with respect to vermiculite and upon further rock weathering the solid phase is formed (tables 30 and 31). At this point a partial equilibrium is maintained between the aqueous solution, vermiculite, Mg-beidellite and kaolinite. At this point the number of independent variables at constant temperature and pressure is 2 and the composition of the aqueous solution will change upon further rock weathering. To maintain the state of partial equilibria between those solid phases and the aqueous solution, magnesium will be consumed from the aqueous solution, the potassium concentrations increased considerably and vermiculite, Mg-beidellite and kaolinite will be formed simultaneously according to (Fe and Al conservation):

$$p(11NaAlSi_3O_8+14KAl_3Si_3O_{10}(OH)_2+3Mg_{4.1}Al_{3.8}Si_{2.1}O_{10}(OH)_8$$

$$+18FeCO_3+3.2Fe_2O_3)+aMg^{2+}+bMgHCO_3^++xH_2O+4.5O_2(g)+yCO_2(g) \rightarrow$$

$$cK^++(11-d)pNa^++dpNaHCO_3+eHCO_3^-+fH_4SiO_4+qMg_{0.167}Al_{2.33}Si_{3.67}$$

$$O_{10}(OH)_2(s)$$

$$+rAl_2Si_2O_5(OH)_4(s)+sK_{0.10}Mg_{6.69}Fe_{0.59}^{III}Al_{1.61}Si_{5.98}O_{20}(OH)_4(s)$$

$$+tFeOOH(s) \tag{7}$$

where $s = (16.9p-1.34q-f)/4.37;$

$t = 24.4p-0.593;$

$r = (64.4p-2.33q-1.61s)/2;$

$a+b = 6.69s+0.167q-12.3p;$

$c = 14p-0.1s;$

$f = 81.3p-3.67q-2r-5.98s;$

$e = (11-d)p+c-2a-b;$

$x = (dp-e+4f+2q+4r+4s+t-52p-b)/2;$ and

$y = dp+e-18p-b.$

Reaction (7) proceeds until the solubility product of muscovite has been reached (points U^1, G^1). The changes in the composition of the aqueous solution and the mass transfer for both partial CO_2 pressures are given in tables 30 and 31. Here again the shifting in the location of the triple point F^1 to G^1 or T^1 to U^1 is too small to be indicated in fig. 30.

From now on rock weathering continues without muscovite of the parent rock being attacked and theoretically it will also be formed mainly because the pH still rises upon further rock weathering. At point U' or G' there is still one independent variable, so the composition of the aqueous solution will be changed and albite and chlorite of the parent rock will be dissolved. The rock weathering theoretically continues until the solubility product of

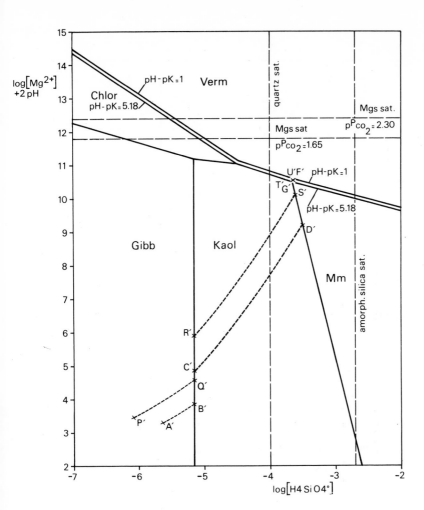

<u>Figure 30</u>   Stability fields and reaction pathways at
P_{CO_2} = $10^{-2 \cdot 30}$ and $10^{-1 \cdot 65}$ bars for the rocks
with a high CO_2 content in the system
$MgO-K_2O-Al_2O_3-Fe_2O_3-SiO_2-H_2O$ $-CO_2(g)$
(298.15°K, 1 bar total pressure). Amorph.
ferric iron governs (Fe^{3+}).
Gibb = gibbsite;   Kaol = kaolinite;
Mm = Mg-beidellite;   Verm = vermiculite;
Chlor = clinochlore;   Mgs = magnesite.

The reaction pathways are P'Q'R'S'T'U' for
P_{CO_2} = $10^{-2 \cdot 30}$ bar;   A'B'C'D'E'F' for
P_{CO_2} = $10^{-1 \cdot 65}$ bar.

Table 28 Mass transfer along the calculated reaction pathways and the composition of the aqueous solution of a rock with a low CO_2 content; $P_{CO_2} = 10^{-2.30}$ bars; $298.15^{\circ}K$ and 1 bar total pressure

Solid phase (moles dissolved, formed or residually enriched by 1 kg of water)

Situation at	Rock dissolved	Goethite formed	Gibbsite formed	Kaolinite formed	Siderite formed and residual	Vermiculite formed	Muscovite formed and residual	Quartz residual	Rutile residual
P	$10^{-8.328}$	$10^{-7.278}$	nil	–	–	–	–	$10^{-5.838}$	$10^{-7.603}$
Q	$10^{-7.388}$	$10^{-6.338}$	$10^{-5.246}$	nil	–	–	–	$10^{-4.898}$	$10^{-6.663}$
R	$10^{-6.415}$	$10^{-5.365}$	–	$10^{-4.526}$	–	–	–	$10^{-3.925}$	$10^{-5.690}$
S	$10^{-5.273}$	$10^{-4.224}$	–	$10^{-3.380}$	nil	–	–	$10^{-2.783}$	$10^{-4.549}$
T	$10^{-5.029}$	$10^{-3.979}$	–	$10^{-3.135}$	$10^{-4.068}$	nil	–	$10^{-2.539}$	$10^{-4.304}$
U	$10^{-4.093}$	$10^{-3.129}$	–	$10^{-2.205}$	$10^{-2.784}$	$10^{-3.534}$	nil	$10^{-1.593}$	$10^{-3.359}$

Liquid phase (moles/kg of water)

Situation at	Mg_t	Fe_t^{II}	K	Na_t	Al_t	H_4SiO_4	HCO_{3t}	pH	pe
P	$10^{-6.947}$	$10^{-7.000}$	$10^{-6.783}$	$10^{-7.137}$	$10^{-6.133}$	$10^{-6.089}$	$10^{-5.744}$	5.150	2.199
Q	$10^{-6.007}$	$10^{-6.061}$	$10^{-5.844}$	$10^{-6.197}$	$10^{-6.422}$	$10^{-5.150}$	$10^{-4.811}$	5.324	0.739
R	$10^{-5.033}$	$10^{-5.087}$	$10^{-4.870}$	$10^{-5.224}$	$10^{-7.027}$	$10^{-6.150}$	$10^{-4.194}$	5.939	-2.069
S	$10^{-3.893}$	$10^{-3.945}$	$10^{-3.729}$	$10^{-4.082}$	$10^{-7.630}$	$10^{-4.052}$	$10^{-3.122}$	7.000	-6.329
T	$10^{-3.647}$	$10^{-4.293}$	$10^{-3.485}$	$10^{-3.839}$	$10^{-7.696}$	$10^{-3.8091}$	$10^{-2.939}$	7.180	-6.507
U	$10^{-4.592}$	$10^{-5.334}$	$10^{-2.543}$	$10^{-2.890}$	$10^{-6.859}$	$10^{-4.063}$	$10^{-2.353}$	7.753	-7.079

Table 29 Mass transfer along the calculated reaction pathways and the composition of the aqueous solution of a rock with a low CO_2 content; $P_{CO_2} = 10^{-1.65}$ bars; 298.15°K and 1 bar total pressure

Solid phase (moles dissolved, formed or residually enriched by 1 kg of water)

Situation at	Rock dissolved	Goethite formed	Gibbsite formed	Kaolinite formed	Siderite formed and residual	Mg-Beidellite formed	Vermiculite formed	Muscovite formed and residual	Quartz residual	Rutile residual
A	$10^{-7.889}$	$10^{-6.840}$	nil	–	–	–	–	–	$10^{-5.399}$	$10^{-7.165}$
B	$10^{-7.385}$	$10^{-6.336}$	$10^{-5.357}$	nil	–	–	–	–	$10^{-4.895}$	$10^{-6.661}$
C	$10^{-6.513}$	$10^{-5.463}$	–	$10^{-4.638}$	–	–	–	–	$10^{-4.023}$	$10^{-5.788}$
D	$10^{-5.067}$	$10^{-4.018}$	–	$10^{-3.174}$	nil	–	–	–	$10^{-2.577}$	$10^{-4.343}$
E	$10^{-4.815}$	$10^{-3.766}$	–	$10^{-2.922}$	$10^{-3.845}$	nil	–	–	$10^{-2.325}$	$10^{-4.091}$
F	$10^{-4.560}$	$10^{-3.510}$	–	$10^{-2.711}$	$10^{-3.394}$	$10^{-3.741}$	nil	–	$10^{-2.070}$	$10^{-3.835}$
G	$10^{-3.764}$	$10^{-2.803}$	–	$10^{-1.887}$	$10^{-2.459}$	$10^{-5.185}$	$10^{-3.222}$	nil	$10^{-1.274}$	$10^{-3.039}$

Liquid phase (moles/kg of water)

Situation at	Mg_t	Fe^{II}_t	K	Na_t	Al_t	H_4SiO_4	HCO_{3t}	pH	pe
A	$10^{-6.509}$	$10^{-6.561}$	$10^{-6.345}$	$10^{-6.699}$	$10^{-5.695}$	$10^{-5.651}$	$10^{-4.621}$	4.871	2.602
B	$10^{-6.004}$	$10^{-6.056}$	$10^{-5.841}$	$10^{-6.195}$	$10^{-5.796}$	$10^{-5.150}$	$10^{-4.552}$	4.935	2.072
C	$10^{-5.131}$	$10^{-5.184}$	$10^{-4.968}$	$10^{-5.322}$	$10^{-6.388}$	$10^{-5.150}$	$10^{-4.183}$	5.304	-0.054
D	$10^{-3.690}$	$10^{-3.748}$	$10^{-3.523}$	$10^{-3.877}$	$10^{-8.222}$	$10^{-3.843}$	$10^{-2.912}$	6.560	-5.213
E	$10^{-3.439}$	$10^{-4.063}$	$10^{-3.271}$	$10^{-8.625}$	$10^{-8.322}$	$10^{-2.594}$	$10^{-2.724}$	6.741	-5.417
F	$10^{-3.208}$	$10^{-4.433}$	$10^{-3.016}$	$10^{-3.369}$	$10^{-8.057}$	$10^{-3.669}$	$10^{-2.518}$	6.939	-5.616
G	$10^{-4.201}$	$10^{-5.372}$	$10^{-2.223}$	$10^{-2.572}$	$10^{-7.565}$	$10^{-3.667}$	$10^{-2.014}$	7.438	-7.014

Table 30 Mass transfer along the calculated reaction pathways and the composition of the aqueous solution of a rock with a high CO_2 content; $P_{CO_2} = 10^{-2.30}$ bars; $P_{O_2} = 10^{-45}$ bars; 298.15°K and 1 bar total pressure

Solid phase (moles dissolved, formed or residually enriched by 1 kg of water)

Situation at	Rock dissolved	Goethite formed	Gibbsite formed	Kaolinite formed	Mg-Beidellite formed	Vermiculite formed	Muscovite formed and residual	Quartz residual	Rutile residual
P^1	$10^{-7.955}$	$10^{-6.568}$	nil	-	-	-	-	$10^{-5.505}$	$10^{-8.557}$
Q^1	$10^{-7.060}$	$10^{-5.673}$	$10^{-5.310}$	nil	-	-	-	$10^{-4.610}$	$10^{-7.662}$
R^1	$10^{-6.424}$	$10^{-5.037}$	-	$10^{-4.929}$	-	-	-	$10^{-3.790}$	$10^{-7.026}$
S^1	$10^{-4.853}$	$10^{-3.466}$	-	$10^{-3.346}$	nil	-	-	$10^{-2.403}$	$10^{-5.455}$
T^1	$10^{-4.686}$	$10^{-3.299}$	-	$10^{-3.264}$	$10^{-3.995}$	nil	-	$10^{-2.236}$	$10^{-5.288}$
U^1	$10^{-3.721}$	$10^{-2.352}$	-	$10^{-2.350}$	$10^{-2.921}$	$10^{-3.497}$	nil	$10^{-1.271}$	$10^{-4.323}$

Liquid phase (moles/kg of water)

Situation at	Mg_t	K	Na_t	Al_t	H_4SiO_4	HCO_{3t}	pH	$p\varepsilon$
P^1	$10^{-6.865}$	$10^{-6.809}$	$10^{-6.914}$	$10^{-6.146}$	$10^{-6.045}$	$10^{-4.991}$	5.144	4.369
Q^1	$10^{-5.970}$	$10^{-5.914}$	$10^{-6.019}$	$10^{-6.362}$	$10^{-5.150}$	$10^{-4.855}$	5.280	4.233
R^1	$10^{-5.333}$	$10^{-5.278}$	$10^{-5.382}$	$10^{-6.735}$	$10^{-5.150}$	$10^{-4.548}$	5.586	3.927
S^1	$10^{-3.760}$	$10^{-3.707}$	$10^{-3.811}$	$10^{-8.081}$	$10^{-3.621}$	$10^{-3.142}$	6.981	2.532
T^1	$10^{-3.622}$	$10^{-3.540}$	$10^{-3.642}$	$10^{-7.891}$	$10^{-3.670}$	$10^{-2.996}$	7.124	2.299
U^1	$10^{-4.901}$	$10^{-2.570}$	$10^{-2.680}$	$10^{-7.225}$	$10^{-3.667}$	$10^{-2.355}$	7.782	1.731

Table 31 Mass transfer along the calculated reaction pathways and the composition of aqueous solution of a rock with a high CO_2 content; $P_{CO_2} = 10^{-1.65}$ bars; $P_{O_2} = 10^{-45}$ bars; 298.15°K and 1 bar total pressure

Solid phase (moles dissolved, formed or residually enriched by 1 kg of water)

Situation at	Rock dissolved	Goethite formed	Gibbsite formed	Kaolinite formed	Mg-Beidellite formed	Vermiculite formed	Muscovite formed and residual	Quartz residual	Rutile residual
A^1	$10^{-7.473}$	$10^{-6.086}$	nil	–	–	–	–	$10^{-5.023}$	$10^{-8.075}$
B^1	$10^{-7.060}$	$10^{-5.673}$	$10^{-5.463}$	nil	–	–	–	$10^{-4.610}$	$10^{-7.662}$
C^1	$10^{-6.537}$	$10^{-5.149}$	–	$10^{-5.082}$	–	–	–	$10^{-4.086}$	$10^{-7.139}$
D^1	$10^{-4.728}$	$10^{-3.341}$	–	$10^{-3.121}$	nil	–	–	$10^{-2.278}$	$10^{-5.330}$
F^1	$10^{-4.199}$	$10^{-2.811}$	–	$10^{-2.840}$	$10^{-3.194}$	nil	–	$10^{-1.748}$	$10^{-4.801}$
G^1	$10^{-3.372}$	$10^{-2.003}$	–	$10^{-2.006}$	$10^{-2.534}$	$10^{-3.157}$	nil	$10^{-0.922}$	$10^{-3.974}$

Liquid phase (moles/kg of water)

Situation at	Mg_t	K	Na_t	Al_t	H_4SiO_4	HCO_3t	pH	$p\varepsilon$
A^1	$10^{-6.383}$	$10^{-6.327}$	$10^{-6.432}$	$10^{-5.664}$	$10^{-5.563}$	$10^{-4.605}$	4.879	4.634
B^1	$10^{-5.970}$	$10^{-5.914}$	$10^{-6.019}$	$10^{-5.774}$	$10^{-5.150}$	$10^{-4.561}$	4.924	4.589
C^1	$10^{-5.447}$	$10^{-5.390}$	$10^{-5.495}$	$10^{-6.029}$	$10^{-5.150}$	$10^{-4.425}$	5.059	4.454
D^1	$10^{-3.368}$	$10^{-3.582}$	$10^{-3.687}$	$10^{-8.644}$	$10^{-3.498}$	$10^{-3.023}$	6.448	3.065
F^1	$10^{-3.173}$	$10^{-3.053}$	$10^{-3.158}$	$10^{-8.071}$	$10^{-3.669}$	$10^{-2.533}$	6.926	2.587
G^1	$10^{-4.221}$	$10^{-2.231}$	$10^{-2.331}$	$10^{-7.522}$	$10^{-3.667}$	$10^{-1.971}$	7.468	2.045

albite or chlorite is surpassed. At that point the
number of degrees of freedom becomes zero and the
aqueous solution must have a constant composition as long
as the rock has not been completely altered. This implies
that overall equilibrium, involving also the parent rock,
cannot be established. This part of the weathering model
was not evaluated because no equilibrium constant of the
specific chlorite of the parent rock is available and it
does not have any practical meaning. In reality muscovite
does not precipitate under standard conditions and the
accompanying theoretical $\lg(K^+)$+pH values are very much
higher than observed in the soil and springwaters from
the Haarts catchment.

3.3.3 Discussion and conclusions

In general the mineral assemblage predicted by the
weathering model for the two rock types for both partial
CO_2 pressures are, with exception of siderite, in close
accordance with the chemical and mineralogical data of the
soil and weathering materials in the Haarts catchment.
However, when the theoretical models are compared with the
field situation in more detail, some limitations are shown
by this theoretical approach.

From the composition of both rock types it can be concluded that as well as a free iron compound, like geothite,
in a closed system (only open for $CO_2(g)$) siderite,
kaolinite and vermiculite or kaolinite, Mg-beidellite and
vermiculite are formed. Mineralogical investigations of
the soil and weathering materials give similar results
(see also chapter B 1.2). The amount of secondary minerals
formed in the weathering models depends on the composition
of the initial mineral assemblage, the environmental
conditions and the resultant weathering paths.
Unfortunately in the Haarts catchment almost no soil
profiles or weathering materials *in situ* have been observed
or the *in situ* formed soil profiles are very young (profile
Haarts 2). Thus a detailed comparison of the amount of new
formed minerals predicted by the weathering models and the
observed mineralogy is hardly possible. Nevertheless some
general conclusions can be made about the predicted
weathering mineral assemblages, but first several assumptions
made in the models will be discussed in more detail.

For both rock types the major component is quartz.
Although the dissolution rate of quartz is very low in
comparison to other minerals in the parent rock, in
practice it releases silica to the aqueous solution. If
quartz contribute to the aqueous silica concentrations in
the soil solution it means that the slopes of the weathering
paths in the gibbsite respectively kaolinite stability
fields will be somewhat less steep and the kaolinite/Mg-
beidellite boundary will be reached at lower $\log(Mg^{2+})$+2pH
(fig. 28 and 30) and $\log(K^+)$+pH values. This implies that
more smectite will be found before the "triple point"
vermiculite/Mg-beidellite/kaolinite will be reached. The
same would apply if a K-mica with a more phengitic

composition (Si:Al > 3:1 in tetrahedral sites was used in the model. The silica concentration levels of a penetrating soil solution can also be, amongst other things, significantly affected within a few minutes by a pH-dependent adsorption/desorption mechanism in soil material (McKeague & Cline 1963a, 1963b). So the initial aqueous silica concentrations can be considerably higher than those of the weathering model in an early stage. Even if it is assumed that rock weathering starts with higher silica concentrations in the lower part in the kaolinite stability field (fig. 28 and 30) a weathering mineral assemblage similar to those predicted by the weathering models will be found. Then if the partial equilibrium concept is valid, due to the ratio Si/Mg/Al in the parent rocks, a simultaneous formation of kaolinite and smectite will always follow and the weathering paths will always arrive at the "triple point" vermiculite/Mg-beidellite/kaolinite.

Theoretical weathering models show of course some limitations which are inherent to the assumptions of the models. The systems presented in the two weathering models are closed with regard to exchange of matter, but not for $CO_2(g)$ and/or $CO_2(g)$ and $O_2(g)$. Both partial pressures are fixed at a constant level. In fact weathering conditions are normally associated with varying P_{CO_2} and P_{O_2} levels in time and depth (Dirven et al 1976). Also an exchange of other matter occurs in nature. This means that a removal of the reaction products and a replacement of the aqueous phase is common. In some cases an open system could be approached if it is assumed that the new formed minerals are not redissolved in the aqueous solution, even if the aqueous solution becomes undersaturated with respect to this specific weathering product. This implies, for both weathering models, that in such an open system the stretches along the gibbsite/kaolinite boundary (BC, QR, B^1C^1 and Q^1R^1) (fig. 28 and 30) do not exist and that the weathering paths will reach the kaolinite/Mg-beidellite boundary at lower $\log(Mg^{2+})+2pH$ and $\log(K^+)+pH$ values. This suggests that all of these paths lead to the vermiculite/Mg-beidellite/kaolinite "triple point" and that also in these "open" systems kaolinite and smectite will be formed simultaneously upon further rock weathering. At this "triple point", in reality the cross section line of the kaolinite, Mg-beidellite and vermiculite spaces, these three solid phases or kaolinite and vermiculite will be formed simultaneously depending on the composition of the two parent rocks.

In the first weathering model (rock with low CO_2 content) the decrease in pε and the appearance of dissolved ferrous iron, siderite and geothite are mainly caused by the decomposition of the ferrous iron-rich chlorites and hematite. Concerning the ferric iron solid phases the Gibbs free energy value of a coarse-grained geothite (Mohr et al 1972) was used in the models. However, at grain sizes below 0.1 µm, mineral solubility is strongly influenced by particle size (Langmuir 1971, 1972; Langmuir & Whittemore

1971) and for particle sizes below 0.07 µm not geothite but hematite is more stable (Van Breemen 1976) and might also be used in the models. Although the solubility of both minerals increases sharply with decreasing particle size, the latter factor or an introducing of hematite instead of geothite does not influence the results of the models due to the inert character of ferric iron. In the observed soil profiles and weathering materials no siderite was detected and in the soil- and springwaters only very low to undetectable ferrous iron concentrations were established. So it might be concluded that weathering proceeds under aerobic or only slightly anaerobic conditions as used in the second model for the rock with the relatively high CO_2 content. Nevertheless, in the first model a closed system with respect to oxygen was used in order to establish extreme conditions and to trace their influences on the formation of secondary silicates. If this is rejected and it is accepted that aerobic or slightly anaerobic conditions prevail, it can be concluded that this assumption does not seriously affect the location of the weathering paths in the MgO - K_2O - Na_2O - Al_2O_3- SiO_2 - H_2O- $CO_2(g)$ subsystem of the first rock type. The slope of the weathering paths in the figures 28 and 29 only very slightly deviates from those presented if all of the iron liberated by rock weathering is fixed as a ferric iron solid phase and gives similar silicate weathering product assemblages. In that case, for the low partial CO_2 pressure also the kaolinite/vermiculite boundary and for the high partial CO_2 pressures the kaolinite/Mg-beidellite boundary is reached at about the same locations indicated by points T and E (fig. 28 and 30).

Various solid phases, like smectite, chlorite, vermiculite and illite show variation in composition and in stability. In the models, Mg-beidellite is considered as a representative of the smectite group of minerals. Montmorillonite gives a less steep field boundary with kaolinite and will generally be formed at higher values of $\log(Mg^{2+})+2pH$, and iron-rich smectites, although not observed in the investigated profiles, occupy about the same field as Mg-beidellite (van Breemen 1976). For vermiculite no other serious thermochemical data are known to the author, so no additional information about the changes in the stability field of vermiculite can be given. It has to be borne in mind that the composition of the vermiculite used in the models deviates from the natural "clay vermiculites" (Weaver & Pollard 1975) and consequently to some extent from their stability fields. The same is true for the fields of chlorites, but this mineral occurs in the weathering materials and soil profiles as an inherited primary chlorite and its stability field will only slightly deviate from that of the clinochlore used in the model. For the K-mica minerals like illite various thermochemical data exist in the literature, but most of these are of rather poor quality and do not always represent real equilibrium solubilities. Therefore only one illite (Helgeson 1968) was considered, but appeared to be metastable with respect to muscovite and is therefore

neglected in the weathering model. However, in spite of the uncertainties of the stability fields of some of the relevant weathering products,
not be affected greatly due to chemical variations in their compositions. Therefore the results of the two weathering models, with the exception of the formation and the residual enrichment of siderite of the first model, will remain unaltered.

The $\log(K^+)+pH$ values in the soil- and springwater and in aqueous solutions which were shaken several months with the parent rocks and weathering materials were clearly lower than those predicted by the weathering models when the solubility product of muscovite is reached. Further, the transformation of the K-mica in the soil profiles also indicates that the points U, U^1, G and G^1 (fig. 28 and 30; tables 28, 29, 30, 31) will never be reached. This suggests that potassium is selectively adsorbed in the soil material, which has often been established in materials containing K-mica and vermiculite, and/or is strongly recycled by the uptakes of potassium by plant roots. Therefore it is believed that the partial equilibria with the appropriate product minerals as kaolinite, smectite and vermiculite or kaolinite and vermiculite or kaolinite and smectite are effective. This indicates that the composition of the aqueous solution during most of the year is near or at the kaolinite/vermiculite boundary or at the "triple point" of kaolinite/smectite/vermiculite boundaries of fig. 28 and 30 (see also chapter B 2.2). As a result of leaching, depending on the relative rates of weathering and water flow, the composition of the soil- and to a lesser extent of the springwaters will be somewhere along the weathering paths presented in the models. Due to the aluminium-rich parent rocks kaolinite will always be precipitated simultaneously with vermiculite and/or smectite, and the composition of the aqueous solution will be somewhere along the planes of the stability space of kaolinite if the weathering rate does not remain too far behind the water flow rate.

4. DISCUSSION AND CONCLUSIONS

Various aspects of weathering and soil formation have been described in previous chapters. In the present chapter an attempt will be made to integrate these aspects to a more complete picture.

All of the rocks in the Haarts catchment belong to the transitional zone, the anchizone, in the high diagenetic/ low metamorphic range. They mostly have a very low CO_2 content and consist of quartz, K-mica, albite, chlorite, with minor amounts of hematite and rutile, with traces of apatite, siderite and pyrite. Also rocks with a relatively high CO_2 content (3.5 - 4 %) have been found, characterised by important amounts of siderite.

Information on weathering processes can be obtained by
studying the mineral assemblages of rock and soils, and
the soil- and springwater chemistry. However, it is
essential first to establish the role of geomorphological
processes which reallocated the parent materials in this
area. In most soils overlying the slope in this area, the
effect of these processes profoundly influences the
alterations caused by pedogenetic processes. In the Haarts
catchment, with the exception of the soils *in situ* on the
narrow plateau tops, all soils have developed in various
slope deposits. At least two, and presumably three, slope
deposits could be distinguished in the watershed: an
older slope deposit, of Pleistocene age, with a grèze-litée
like habitus. This deposit is only present on the steep
lower slopes. Although thick layers of this stratified
slope deposit are known in the Oesling, they are of limited
importance in the Haarts catchment due to their shallowness.
It is covered by an intermediate slope deposit of presumably
late Würm or in any case pre-Allerød age, characterised by
an absence of volcanic minerals of the Laacher See eruption
of the Allerød, a high gravel and stone content, without
pollen and charcoal but with a truncated soil profile comprising an argillic horizon. The pedological processes
responsible for the formation of this soil have been
brought into relation with the conditions of slope stability
prevailing in this area during much of the Holocene up to
the time of the large-scale deforestation in the Middle
Ages. This soil was eroded and subsequently covered by:
a young slope deposit formed in the period from 1450 to
1800 AD. It is characterized by a specific pollen stratification, a high charcoal content, volcanic minerals and a
cambic horizon (Umbric Dystrochrepts). The soil represents
the period of landscape stability under forest since ca.
1800 AD. Due to the intricate history of weathering,
erosion, deposition and soil formation, the clay mineralogy
of the area shows variations which will be discussed below.

4.1   The history of weathering

The history of weathering begins with the formation of
the grèze-litée. Like its counterparts in southwestern
France, this layered slope deposit was apparently formed
purely by physical disintegration of the shales to
individual shards, without concomittant chemical changes.
The periglacial environment of the late Pleistocene seems
to offer the most likely conditions for this type of
weathering. The same conclusion can be drawn for the
intermediate slope deposit. Mass movements like solifluction
and other related slope forming processes in a periglacial
environment must be responsible for the deposition of these
materials on the slopes.

The beginning of the chemical attack on the unweathered
rocks and the Pleistocene deposits is difficult to establish.
The grey rocks are readily coloured brown upon weathering,
indicating that ferrous iron is oxidized and fixed as ferric
iron. This means that in rocks with a relatively high CO_2
content, siderite must be transformed mainly in geothite.

In all rock types a strong relationship exists between K_2O and MgO content and, for the rocks with a low CO_2 content, also between the K_2O and MgO+FeO content, probably suggesting that K-mica has some chloritic layers (phengitic composition). These chloritic layers, as well as the mixed-layer "illite"/chlorite and the chlorite presumably have a considerable amount of ferrous iron, which appears to be common in some low-grade metamorphic rocks.

As soon as this rock was affected by weathering, the composition of the remaining rock altered considerably. The MgO content was lowered and no positive relationship exists now between the K_2O and MgO and/or between the K_2O and MgO+FeO content. The ferrous iron content of these weathered rocks is very low. Obviously a considerable part of the chlorite and the chloritic layers which are rich in ferrous iron have been transformed. Their ferrous iron was oxidized and fixed as geothite. This conclusion is corroborated by the low ferrous iron content of the fine earth in the subsoil horizons and in the gravel and stones of the Haarts catchment.

From a mineralogical point of view, the effect of weathering is limited. The fractions < 2 μm of powdered specimen of both fresh and slightly weathered hard rock consists of K-mica (illite), with only small portions of quartz, chlorite, a mixed-layer K-mica/chlorite and hematite. The clay mineralogy of the grèze-litée and the other pre-Allerød slope deposits show great similarity with that of the powdered rock specimen. K-mica (illite) clearly remains dominant, and quartz, chlorite, hematite and mixed-layer K-mica/chlorite are all present in sub-ordinate amounts. Weathering results in the disintegration of gravel and stones. However, not all of the somewhat higher content of fine material in the intermediate slope deposit can be attributed to weathering *in situ*. It is clear from the difference in texture with the underlying grèze-litée that some of the fines must be derived from the site of origin of the intermediate slope deposit. This means that the disintegration of the rocks antedates the formation of this deposit.

The intermediate slope deposit underwent soil formation in Holocene times resulting in an argillic horizon of which remnants are still found on the lower slopes. The sand and silt fraction of this soil show only gradual changes in respect to the composition of the "weathered" rock. It could be concluded from the decrease of the potassium content of these fractions and from the X-ray analyses that K-mica is effectively attacked by weathering whereas quartz is residually enriched. Also the albite content slightly decreases. The magnesium and ferrous iron content of these fractions are also more or less the same within the distinguished slope deposits. This and the decreasing K_2O/ MgO ratios towards the surface suggest that the chlorite which is low in ferrous iron is only slightly altered upon weathering. In contrast increasing or constant K_2O/MgO ratios were established for the silt and sand fractions of the slope deposits containing relatively high ferrous iron chlorite, indicating that this chlorite is transformed at least to the same extent as illite by weathering.

The soil formation (Dystrochrept) affecting the young slope deposit and the young soil on the narrow plateau top represents the recent phase of weathering. In the clay fractions a relatively high amount of vermiculite and also an Al-hydroxy interlayered 14 Å mineral (no K contraction, 300°C heating shoulder 10.5-13.5 Å 500°C heating Å peak) were detected. This pedogenic 2:1 -2:2 mineral was interpreted as an Al-hydroxy vermiculite, although the presence of an Al-hydroxy interlayered smectite cannot be excluded. From the decrease of the 14 Å peak upon K saturation and the increase of the 10 Å peak and also of the very small 17 - 17.6 Å peaks upon ethylene glycol treatment, it has been concluded that most of the 14 Å peak of the Mg-saturated clays represents vermiculite. It dominates the Al-hydroxy interlayered vermiculite in the lower part of the Dystrochrepts, although the latter increases towards the surface and may even dominate vermiculite in the higher cambic horizons (profiles Haarts 1, Scheissgrond 1). However, in the topmost Al horizons of the Dystrochrepts, and in the young soils of the plateau tops, vermiculite is always more abundant than the Al-hydroxy interlayered mineral. This can be explained by complexing of the aluminium by organic substances resulting in a minor filling in of the interlayer positions of the vermiculite (Rich 1968). Also traces of smectite (17 - 17.6 Å spacing upon ethylene glycol treatment) could be observed mainly in the surface soil. In the diffracto-grams 16 Å peaks appear upon ethylene glycol treatment. These peaks can be interpreted as vermiculite because clay vermiculite can expand to 16 - 16.5 Å in the presence of ethylene glycol (Walker 1975), or as a mixed-layer vermiculite-montmorillonite. Kaolinite could be detected with the help of the 060 reflections, the intensity ratio of the 7 Å/14 Å peaks and the dimethylsylfoxide treatment (11.2 Å spacing).

The silt and sand fraction of the youngest slope deposit and of the young soils show similar features as those described for the intermediate slope deposit. The youngest slope deposit has a somewhat higher ferrous iron content, which might be caused by the presence of vermiculite and/or (Al+FeII)) hydroxy interlayered mineral (Carstea *et al* 1970a, 1970b) in the silt fraction (Coffmann & Fanning 1975).

Some of the changes described for the young soils and young slope deposits are also noted in the upper part of the intermediate slope deposit. Little vermiculite could be detected without any sign of Al-hydroxy interlayering. However, these alterations probably do not date from the time of the Early Holocene soil formation which lead to the argillic horizon now found in truncated soils. Apparently such alterations as then affected the upper hori-zons of this soil disappeared by erosion before the young slope deposit was laid down. It is the effect of the present-day weathering which reached the intermediate slope deposits especially where the cover of the young slope deposits is thin, ie. on the upper slopes. Here more vermiculite could be detected than in the intermediate slope deposit on the lower slopes. Also smectite is present in minor quantities, as is Al-hydroxy interlayered vermiculite which is absent lower in the profile.

From the previous section it is clear that the clay mineralogical composition of the various soil horizons cannot be explained by only considering weathering processes. No complete mineralogical weathering sequence could be detected in the clay fractions of the successive soil horizons mainly due to the mixed character of the weathering materials (slope deposits). Nevertheless it could be established that weathering processes in the Haarts catchment have resulted in the transformation of dioctahedral K-mica (illite) into vermiculite by potassium replacement from the interlayer positions and that kaolinite has been formed. The rather low kaolinite contents, if the present soil- and springwater chemistry is considered, probably indicates slow kinetics for kaolinite precipitation (assuming equal environmental conditions in the past). If the chlorite content of the bedrock is compared with that of the weathered rock, the chlorite rich in ferrous iron must have been transformed. The possible contribution of this mineral to the formation of vermiculite could not be evaluated by chemical- and X-ray analysis due to the complicated erosion and deposition processes and to the low chlorite content of the weathering materials. Transformation of relatively iron-rich chlorites into vermiculites by structural desorganisation of the hydroxide sheet by dehydroxylation and oxidation of ferrous iron has often been reported in the literature (Ross 1968, 1969; Ross & Kodama 1974; Adams 1976). The Mg-chlorite low in ferrous iron present in the weathered rock and in the weathering materials of the intermediate slope deposit seems to be at most slightly altered (14 Å reflection upon heating to 550°C throughout the profile), a fact which is supported by experimental weathering data (Ross 1975). Therefore it is believed that the bulk of the vermiculite is coming from weathering of K-mica (illite).

The formation of the Dystrochrepts in the young slope deposits is accompanied by the formation of vermiculite, kaolinite and Al-hydroxy interlayered mineral. Also minor amounts of smectite, primary chlorite and perhaps a mixed-layer vermiculite-smectite (16 Å peak upon ethylene glycol) have been observed, but almost no real "soil chlorites" could be established. Obviously the time of pedogenesis is still too short or the environmental conditions are unfavourable (high organic matter content) to create soil chlorites (completely chloritized vermiculites). Although all of these minerals could be (partly) inherited from an older soil forming period (original surface soil material of the truncated soil with the argillic horizon), it could be established that most of these minerals with the exception of "primary" chlorites have been formed by weathering since 1800 AD and (partly) represent the present-day environmental conditions (see also spring- and soil water chemistry). So

it can be concluded that part of the vermiculite formed is
altered into an Al-hydroxy interlayered vermiculite and
perhaps also into mixed-layer vermiculite-smectite and
smectite. The Al-hydroxy interlayering is probably
favoured because the aluminium and iron mobilization by
chelating is limited to the upper soil horizons (fig. 26).
At a certain depth microbiological oxidation of the organic
acids probably plays an important role, resulting in
locally high Al^{3+} activity and formation of Al-interlayered
minerals.

The formation of vermiculite and Al-hydroxy interlayered
minerals is also reflected in the CEC capacity of the soils,
and of the clay fractions. If the influence of organic
matter is left out, the CEC of the clay fractions increases
towards the top of the intermediate deposit, suggesting
vermiculite formation. In the Dystrochrept, the CEC of the
non-organic clay fraction decreases in the same direction
due to the formation of the Al-hydroxy interlayered
vermiculite. The clay fractions in the argillic horizon
contain considerably less ferrous iron than those of the
Dystrochrept. This difference is probably caused by the
Al-hydroxy interlayering, whereby ferrous iron can be
built into the interlayer positions (Brinkman 1977).

Atmospheric precipitation-, soil- and springwater
chemistry and the theoretical weathering model

In the Haarts catchment there is no evidence of any groundwater contribution from outside the drainage basin. This
can be concluded, amongst other things, from the relatively
high position of the springs in the catchment compared with
those of the adjacent watersheds, and from the absence of
a deep groundwater reservoir (see A II 3). Also the low
nitrate content of the spring- and riverwater in the forested
Haarts catchment is striking, indicating that all water
is from local origin. The spring- and riverwaters of the
surrounding drainage basins, all with a considerable part
of arable land, have much higher nitrate levels. No
quantitatively important water leakage by deep seepage is
expected as the joints in the bedrock are closed within a
few meters depth (A II 3). This is supported by the water
balance and the chloride budget (A II 1: Imeson &
Verstraten 1979). Because no change in storage has been
observed at the beginning and the end of the two hydrological
years November 1973-October 1975, the water balance equation
can be simplified to: Precipitation = Evapotranspiration
+ Discharge. If these quantities are taken into account,
it can be concluded that water leakage is negligible. The
rocks do not contain chlorides, and chloride is not
accumulated in important quantities in the catchment.
Therefore if losses by deep seepage are negligible, then
chloride inputs by precipitation equals chloride output in
stream water (Likens et al 1977). Outputs of chloride are
not significantly different from inputs in the Haarts
catchment (Verstraten 1977). So it can be concluded that
the watershed is watertight.

From hydrograph analyses and field measurements it can be established that in the saprolite water percolates predominantly downwards with little lateral movement. To a depth of ca. 2.5 m throughflow in these permeable materials was insignificantly small. So it can be assumed that lateral transport of water to the valley bottom takes place in the grèze-litée like slope deposits and probably in the very shallow joints in the uppermost part of the Lower Devonian rocks. This is corroborated by the annual range in temperature of the springwaters mentioned earlier. Therefore it can be concluded that the chemical composition of the springwaters must be strongly related to the weathering processes.

Although the chemical composition of the atmospheric precipitation in the Haarts catchment is strongly influenced by industrial pollution, its contribution to the major elements (Mg, Ca, Na and K) which are liberated by weathering can be almost completely ignored (Verstraten 1977). Only chloride, nitrogen and sulphate are greatly affected by the input of the precipitation, but with the exception of sulphur these elements are not present in the parent rocks. However, the acidity of the precipitation caused by the above mentioned elements is of great importance for the chemical weathering processes. As in all industrial areas, the driving force of the chemical weathering, the hydrogen ion, is supplied from two sources. One source is external to the ecosystem and the other internal (Likens *et al* 1977). The external source is the input of hydrogen ions (acidity) by bulk precipitation. As discussed in part B 2.1 the average input of hydrogen ions (free acidity) derived from sulphuric, nitric and hydrochloric acid was 0.3×10^3 eq/ha.yr (table 10) during the investigation period. In addition to this, the average contribution by carbonic acid in the precipitation to the input of hydrogen ions was 0.2×10^3 eq/ha.yr. Actually this contribution will be somewhat lower because the pH of the weathering environment is lower (pH < 7.4) than the pH necessary for a complete dissociation of carbonic acid. On the other hand, the mean annual output for metal cations, assuming steady state conditions in the catchment was 2.85×10^3 eq/ha.yr (Verstraten 1977). This implicates, if hydrogen is completely consumed by chemical weathering reactions, that another more important hydrogen source, an internal one, is operating effectively. If no steady state conditions prevail and for instance biomass production acts as an important long-term sink for several cations, the internal hydrogen sources are even more important, because then the chemical denudation rate has been underestimated.

According to Likens *et al* (1977), "internal" means everything that is not sensibly added by meteorological input. Sulphur, ammonium, carbon and aluminium and hydrogen at the soil adsorption complex are at first sight the most important internal hydrogen sources and will be briefly discussed.

As the low-grade metamorphic rocks and the weathering material contain traces of pyrite, the sulphate content of

the soil - and especially the springwaters might at first
thought to be controlled by pyrite dissolution. However,
the fluctuations of the sulphate content of the springwaters throughout the year is such that this possibility
has to be rejected. A slightly negative relationship
(r = -0.20) exists between the sulphate and the alkalinity,
the latter being to some extent an overall measure for
chemical weathering processes in the catchment. So it can
be concluded that pyrite oxidation and microbiological
decomposition of organic sulphur compounds in the soils and
the resulting hydrogen ion production play at most a very
subordinate role. This is confirmed by the input-output
data of sulphur in the catchment (Verstraten 1977).

Ammonium is added to the weathering system by bulk
precipitation input in considerable amounts (0.4×10^3 eq/
ha.yr) and by the decomposition of organic matter within
this system. If complete oxidation of ammonium to nitrate
takes place, two hydrogen ions are produced. However, most
of the ammonium in the soil environment is fixed by plants
and only about 10 to 20% will be oxidized (Melillo 1977).
Nevertheless, nitrification is an additional source of
hydrogen ion production, although it seems to be somewhat
limited.

Organic matter decomposition involves partial or complete
oxidation, resulting in the production of carbon dioxide and
water or intermediate products such as soluble organic
acids. Partial CO_2 pressures between $10^{-1\cdot65}$ to $10^{-2\cdot30}$
bar have been established in the catchment resulting in a
hydrogen ion production varying between 6.3×10^3 and
1.4×10^3 eq/ha.yr ("effective" precipitation = 797 mm for
the investigated period, table 3). Also the soluble organic
acids like low molecular aliphatic and aromatic acids, as
well as "fulvic" and "humic" acids containing carboxylic
and phenolic groups, contribute to the hydrogen ion
production, although no quantitative data can be presented.

Finally the inorganic part of the soil adsorption complex can also contribute to the hydrogen ion production or
the hydrogen ion uptake. Hydrogen ion production is mainly
based on the deprotonization reactions of adsorbed
aluminium. Although aluminium is the main cation at the
adsorption complex of the Dystrochrepts and the Udorthents,
its influence on the hydrogen production seems to be limited
due to the low pH of the soil waters.

Thus it can be concluded that hydrogen ion production in
the Haarts catchment is mainly governed by the mineralization
of organic matter. This is also confirmed by the highly
positive significant correlations between the main cations
and the alkalinity of the springwaters. The hydrogen input
by polluted precipitation cannot be completely ignored, but
is of minor importance.

If the water enters the soil chemical weathering takes
place through the hydrogen input. The chemical weathering
reactions, if the upper part of the soil system with the
Al-hydroxy interlayering is excluded, seem to be simple
dissolution-precipitation reactions. The main rock con-

stituents albite, chlorite, K-mica and siderite are attacked, resulting in an incongruent dissolution reaction and formation of kaolinite, ferric iron oxide (geothite) and release of sodium, magnesium, potassium, aqueous silica and bicarbonate. The composition of the springwaters indicate that this incongruent dissolution reaction (kaolinitization) seems to be effective during most of the year. However, at low flow conditions (summer periods) with alkalinities above 1.2 meq/l, partial equilibria seem to exist between kaolinite and vermiculite and/or between kaolinite and smectite (fig. 23 and 24). These partial equilibria have been evaluated from the curves of the plotted values of the disequilibrium indices. From mineralogical investigations (see B 1.2) it seems reasonable to conclude that especially the kaolinite-vermiculite partial equilibrium is effective.

Mass balance calculations (table 21, 22) indicate that particularly albite and chlorite rich in ferrous iron contribute to the release of cations and aqueous silica in the springwaters and to the formation of kaolinite. Also apatite dissolution seems to be relatively important if the relatively high calcium concentration levels in the spring-waters are considered. Phosphor, liberated by the apatite dissolution, is apparently fixed in the weathering system, because only very low orthophosphate concentrations have been established in the spring- and riverwaters in the Haarts catchment. The precise contribution of K-mica to the release of cations, aqueous silica and weathering products cannot be evaluated accurately due to the selective uptake of potassium from the soil- and spring-waters by plant roots and/or adsorption. This results in a contribution of K-mica in the mass balance calculations which is too low (table 21, 22). Nevertheless it can be concluded that most rock constituents are affected by weathering in the kaolinitization stage; in the vermiculitization (smectitization)-kaolinitization stage clay minerals and ferric iron oxides are formed but to a much lesser extent (table 22).

The results of the theoretical mass transfer calculations compare favourably with the (clay) mineralogy of the saprolites. The theoretical weathering models predict a similar mineral assemblage if the mass transfer calculations are continued until points somewhere between T and U, F and G, T and U^1 and F^1 and G^1 (fig. 28 & 30) have been reached. The theoretical weathering model predicts a kaolinite, vermiculite and/or smectite and ferric iron assemblage, especially when the redox potential of the weathering system is governed by the partial oxygen pressure and not by the ferrous/ferric iron couple. The predicted formation of kaolinite and the simultaneous precipitation of kaolinite, vermiculite and/or smectite have been confirmed by the springwater chemistry. The latter also indicates a formation of kaolinite and a partial equilibrium between kaolinite, vermiculite and/or smectite and consequently a simultaneous formation of these minerals.

However, when the weathering paths of the theoretical model are compared with the composition of the soil- and

149

springwaters some remarks should be made. The soilwater
(porous cup) and springwater compositions indicate a steeper
slope of the weathering paths in the kaolinite stability
field than those predicted by the mass transfer calculations
(fig. 22a, 28 & 30). The divergence can be explained by
different relative reaction rates of the rock forming
minerals, which are unlike those used in the model. For
instance, if the reaction rates of these minerals as
established by the mass balance calculations (table 21 &
22) had been used, steeper weathering paths in the kaolinite
stability field would be obtained. This can easily be
shown if the incongruent dissolution reaction rock →
kaolinite is considered. From the springwaters the apparent
average relative reaction rates for albite, K-mica and
$Mg-Fe^{II}$ chlorite are 21:2:17.9 and result in:

$$(NaAlSi_3O_8+2KAl_3Si_3O_{10}(OH)_2+$$

$$17.9Mg_{2.4}Fe_{1.9}^{II}Al_{3.6}Si_{2.15}O_{10}(OH)_8)(s)+(109-a-b)CO_2(g)+xH_2O \rightarrow$$

$$(21-a)Na^+ + aNaHCO_3^0 + 2K^+ + (43-b)Mg^{2+} + bMgHCO_3^+ + 34FeOOH(s) +$$

$$45.7Al_2Si_2O_5(OH)_4(s) + 16H_4SiO_4^0 + (109-a-b)HCO_3^-$$

If the coefficient of magnesium and aqueous silica of this
reaction are compared with those in the theoretical weathering
model the alternative weathering paths will be steeper than
those of the model and in closer accordance with the plotted
composition of the soil- and springwaters. A complete
quantitative evaluation of these alternative dissolution
reactions will be presented elsewhere (Verstraten 1979a).
Nevertheless the conclusion given by the theoretical model
on a simultaneous formation of kaolinite, vermiculite and/
or smectite can be maintained with these alternative
reaction rates.

4.3 Chemical denudation in the Haarts catchment

Finally an attempt will be made to calculate the
chemical denudation rate in the forested Haarts catchment.
Only the constituents which can be derived from the rocks
are taken into account for this calculation (table 33).
A knowledge of the precipitation and runoff chemistry is
essential for the calculation of this parameter.

4.3.1 Water composition

Precipitation. Precipitation is the only source of
water entering the drainage basin and carries about 23 per
cent of the relevant solutes (SiO₂, K, Na, Ca, Mg)
eventually discharged in the runoff. The weighted mean
ionic concentration of this dissolved matter in the
precipitation is 1.8 mg/l (table 32). No seasonal trends
in the ionic concentrations could be detected. Chloride
and sulphate are the major anions, sodium, calcium and
ammonium the principal cations (table 9). Precipitation in
the catchment is distinctly acid (pH 4.4) and the hydrogen

Table 32   Weighted mean ionic concentrations (mg/l) of
            precipitation and stream waters[*]

	SiO$_2$ (aq)	K	Na	NH$_4$	Ca	Mg	HCO$_3$	Cl	NO$_3$	SO$_4$	H	pH
Precipitation	0.1	0.25	0.67	0.87	0.74	0.07	-	1.93	1.62	3.94	0.04	4.4
Stream water "summer"	7.8	1.0	5.3	0.02	6.45	8.1	54.4	6.1	2.9	6.0	tr	7.3
Stream water "winter" and "traditional"	7.1	0.5	4.0	0.02	3.65	4.2	12.1	6.2	6.6	6.15	tr	6.7

[*] November 1973 - October 1975

input contributes to the weathering of the shales. The
input of hydrogen ions by precipitation in the watershed
is about 10 kg, but only 0.03 kg of hydrogen ions is
removed in the runoff.

   Runoff.

      Base flow.   About 40 per cent of the water dis-
charged from the Haarts watershed could be considered as
base flow. As the water composition of the delayed flow
during the "summer" period differs significantly from that
during the remainder of the year it is considered separately.
The "summer" delayed flow comprised about 4 per cent of the
total runoff and carried approximately 10 per cent of the
total dissolved matter output from the drainage basin. The
chemical composition of this base flow is a function of the
residence time of the water in the weathering materials,
and the total dissolved matter varied from the beginning
until the end of the summer, from 77 mg/l to 145 mg/l.
Bicarbonate, magnesium, calcium and aqueous silica formed
the highest percentual content of the total dissolved matter
in their base flow component.

   Base flow during the remainder of the year contributed
about 36 per cent of the total runoff. Its total dissolved
matter content varied from 50 to 62 mg/l. Most striking
in comparison with the "summer" base flow is the low
bicarbonate concentration. Bicarbonate, chloride, magnesium,
calcium, sodium, nitrate and aqueous silica are the main
ions.

      Quick flow.   The quick flow in the "summer" period is
of minor (1.5 per cent) importance to the total runoff and
its contribution to the output of the dissolved matter is
less than 0.5 per cent. In particular concentrations of
mineral matter derived from rock weathering (magnesium,
calcium, sodium, silica and bicarbonate) show a strong
decrease in the "summer" quick flow. Potassium, chloride and
sulphate concentrations are approximately the same or even
increase, due to the relatively important contribution of

151

"channel precipitation" during this period. The quick flow
in the "winter" and "transitional" periods formed by the
broad hydrograph peaks comprised about 58 per cent of the
total runoff. The chemical composition of these quick
flow components during these two periods is not significantly
different from each other. The contribution of the "first
peak"-quick flow in the "transitional" period (spring and
autumn) is negligible in terms of the dissolved matter output of the total quick flow component. In spite of the
large number of water samples taken from base flow and
floods in the "non-summer" periods no characteristic
chemical composition for the corresponding water components
(see A II 5.5) could be detected. If there is a difference
in composition between these "corresponding" components for
the two periods, it will be small, complex, and only made
apparent by the continuous recording of the electrical
conductivity and the activities of the various ions. This
suggests that although the "non-summer" runoff can be considered as being supplied from a number of reservoirs with
distinct reaction factors (see A II 5.5.4), it does not
necessarily have any significance for the water chemistry.
The similarity of base and quick flow water chemistry meant
that the output of solutes during the "non-summer" period
could simply be calculated from a weighted mean ionic
concentration obtained from the considerable number of samples
representative for these periods (table 32). Bicarbonate,
aqueous silica, chloride, nitrate,sulphate, magnesium and
calcium are the principal ions, but only those species which
are derived from the rocks are taken into account for the
chemical denudation rate.

      In general the quantity of dissolved matter removed from
the catchment is closely related to the total volume of the
flow. The volume of flow during the "summer"period over the
two years observations was 7,340 m^3 (5.7 per cent of the
total runoff) and this carried 750 kg (10.1 per cent of
the total output of solutes) dissolved matter out of the
drainage basin. The volume of flow during the rest of
the year was 122,260 m^3(94.3 per cent) and this transported
6,698 kg (89.9 per cent) of material. The concentrations
of the dissolved matter, however, are affected amongst
other things by residence time of the soil waters,
precipitation patterns and the soil moisture regime.

4.3.2   Chemical denudation rate

      In order to calculate the chemical denudation rate
from the input and output data three assumptions were made
in this study: first,that all waters draining the watershed originate as precipitation within the watershed;
second, that the biomass is in equilibrium; and third, that
the storage of water in the saprolite in November 1973 was
the same as in November 1975. The last assumption was confirmed by soil moisture measurements which indicated that
no significant changes in the volume of water in the saprolite had occurred. The stability of the biomass might be
suggested by the following evidence:

1. Trees of all size classes from first-year sapling to old high trees occur in the catchment

2. Dead falls are present in the watershed.

3. Indications for logging during the last 15 years are absent.

There is no evidence of any groundwater contributions to the stream from outside the drainage basin and sufficient water can be stored in the thick soil and weathering material to maintain base flow in the stream (Imeson & Verstraten 1979).

The weathering of silicate minerals and apatite in the drainage basin accounts for silica, alkali and alkaline earth elements in excess amounts introduced by rain. Silica, calcium, sodium and magnesium are assumed to be in equilibrium with the biomass. Sodium and magnesium are not major plant nutrients, so they will be minimally affected by the biomass. In the calculations the quantity of calcium introduced by chemical weathering of apatite has been obtained by substracting input from output data. As calcium is a major plant nutrient this value might be a minimum one, and the contribution of the chemical weathering to the calcium output might be somewhat greater. Because almost all phosphorus species are fixed in the biomass or by adsorption in the saprolite verification is not possible. The effect of chemical weathering to the output of potassium is somewhat questionable. From the analyses of soil water from various soil horizons it is quite clear that potassium has been continuously recycled by the biomass. The potassium concentrations decrease downwards from the A1 to the B3 and R1 horizon (fig. 26). From the norm calculations and the X-ray analyses (see B 1.2) it is concluded that the micaceous minerals are attacked in the saprolite. It must therefore be expected that potassium liberated by chemical weathering will be fixed by adsorption, or that it is partly picked up by the biomass which is a state of chemical disequilibrium with respect to this element. In both cases the contribution of the chemical weathering to the potassium output is also a minimum quantity.

In the Haarts catchment the output of solutes to which weathering of the rocks contributes was:

SiO_2 27.6 kg/ha.yr; K 2.0 kg/ha. yr; Na 15.6 kg/ha.yr;

Ca 14.6 kg/ha. yr; Mg 16.9 kg/ha. yr.

This means that in the period November 1973 - October 1975 the output of the dissolved matter was 2593 kg (76.7 kg/ha. yr) (table 33). If steady state conditions prevail and consequently biomass does not operate as a "long-term sink" for some of the nutrients provided by chemical weathering, the actual chemical denudation rate in the Haarts catchment can be established by subtracking the input from the output data. However, this chemical denudation rate will be a

153

Table 33   Input, output and chemical denudation rate
(in kg) for the rock constituents in the
Haarts catchment (16.9 ha) [x]

	SiO_2 (aq)	K	Na	Ca	Mg	Total	SO_4
Input (by precipitation)	16	68	182	201	20	487	1070
Output "summer"	58	7	39	47	60	211	50
Output "winter + transitional"	874	62	489	446	511	2382	751
Output total	932	69	528	493	571	2593	791
Chemical denudation (weathering)	916	1	346	292	551	2106	-

[x] November 1973 - October 1975

minimum value, mainly due to the uncertainties in the
steady state conditions of the biomass.   In the Hubbard
Brook Experimental Forest, for instance, it could be
established that elements provided by chemical weathering
were still stored in the biomass after a logging operation
55 years ago (Likens *et al* 1977).   This effect would
result in this ecosystem in an underestimation of the
cationic chemical denudation rate (net loss from ecosystem
plus long-term storage within the system) of 50-60 percent
(based on eq/ha.yr).   However in the Haarts catchment logging
was of minor importance during the last 60 years.
Therefore it is believed that the effect of element storage
in biomass plays at most a very subordinate role.   The
chemical denudation in the Haarts catchment for the
investigated period amounted to 2106 kg (table 33), which
is a rate of 62.3 kg/ha. yr and the contribution of the
various elements to this parameter was:

SiO_2 27.1kg/ha.yr;   K 0.0 kg/ha.yr;   Na 10.2 kg/ha.yr;
Ca 8.6 kg/ha. yr;   Mg 16.3 kg/ha.yr.

These values indicate that the contribution of the chemical
weathering to the output of the dissolved matter which
could be derived from rocks, was 81.2 per cent.   These
data also indicate that the output of material in solution
in roughly 4 times greater than the output of material in
suspension and as bedload (see also Imeson 1976).   The
average lowering of the rock in the catchment by chemical
denudation amounts to approximately 0.5 mm in the last

154

200 years. This lowering is based on the assumption that the specific weight of the rock is 2.5 g/cm^3

The cationic denudation in the Haarts catchment allows to estimate the relative importance of the external (meteorological origin) and internal (soil environment) sources of hydrogen ions as weathering agents. The cationic denudation was estimated as 2.22 x 10^3 eq/ha.yr (table 33), which should be balanced by the sum of the net external and net internal supply of hydrogen ions. The external supply rate is about 0.5 - 0.6 eq/ha. yr, if about 20 % of all ammonium is oxidized to nitrate. Consequently by substracting the internal sources for hydrogen supply 1.6 - 1.7 X 10^3 eq/ha.yr. This value suggests that about 75 percent of the hydrogen is coming from the internal source, the soil environment and is almost completely provided by mineralization of organic matter.

SUMMARY

This study presents the results of an investigation of the weathering processes in (very) low-grade metamorphic rocks and the soil formation in the weathering materials in a completely forested and unhabited catchment in the Luxembourg Ardennes (fig. 1).

In this study on soil genesis not only the solid phases but also the liquid phase was investigated. For over two years, May 1973 - October 1975, this liquid phase (soil-, spring and river waters) were periodically sampled. Due to the large mass ratio of solids to solutes for most elements in the weathering system, undetectable small changes in the composition of the rocks and soils show up as easily detectable, relatively large changes in the composition of the liquid phase. This will be especially the case in relatively young soils like those in the Haarts catchment. If the concentrations of solutes are recalculated into activities, the ion activity products for the relevant solid phases can be evaluated in order to establish to what extent the solution is in equilibrium with specific minerals. This approach results in a better understanding of the weathering processes.

In the first part of this study the physiographic factors are discussed which played an important role during the Late Quaternary for the weathering materials.

The Haarts catchment is situated somewhat south of the central part of the Wiltz synclinorium (fig. 5). Here a rather simple symmetrical structure exists, resulting in a generally undisturbed sequence of formations. The catchment consists of Lower Devonian rocks which all belong to the Anchizone, the transitional zone between the high-diagenetic/low-grade metamorphic range. According to Lucius's (chrono) stratigraphic division the catchment consists of the "Schiefer von Stolzemburg" with only in the northern part a small area of "Quartzophylladen von

Schüttburg" are characterised by frequent intercalations
of quite massive dark brownish quartz-sandstone, which
occurs interbedded with massive well-compacted phyllites
and quartzphyllites. Joints occur frequently, although
their influence is very restricted at some depths. The
joints are mostly closed within one metre, resulting in an
extremely small water reservoir for the Lower Devonian
rocks, in which water circulation is only possible within
the few upper metres of these rocks.

The Haarts catchment is morphologically a part of a
gently undulating planation surface with steeply incised
valleys. Four geomprhic units could be distinguished
(fig. 11): a. small plateau tops; b. gentle upper
slope; c. steep lower slope; d. dry valley.

With the exception of the small plateau tops the whole
area is covered with slope deposits. Three slope deposits
could be detected with the help of palynological,
mineralogical and micromorphological investigations:

- an older slope deposit of Pleistocene age with a
  grèze-litée like habitus. This deposit is only present
  in the steep lower slopes and is of limited importance
  due to its shallowness (20-40 cm).

- an intermediate one of presumably late Würm or in any
  case pre-Allerød age, characterised by an absence of
  volcanic minerals of the Laacher See eruption of the
  Allerød, a high gravel and stone content, without
  pollen and charcoal, but with a truncated soil profile
  comprising an argillic horizon. The soil forming
  processes for this soil with an argillic horizon took
  place during much of the Holocene up to the time of
  large-scale deforestation in the Middle Ages.

- a young slope deposit formed in the period from 1650 -
  1800 AD. It is characterised by a specific pollen
  stratification, a high charcoal content, volcanic
  minerals and a cambic horizon. Only in the upper dry
  valley in this deposit no cambic horizon was developed.

Due to the changing environmental conditions during the
Late Quaternary and the resultant erosion, deposition and
soil formation, the soils in the Haarts catchment,with the
exception of those of the small plateau tops, are polygenetic.

- On the small plateau tops thin soils occur with an
  A1-R or an A11-A12-R profile (Lithic Udorthent and
  Lithic Umbric Dystrochrept)

- On the valley slopes soils with an argillic horizon
  occur, covered by a young slope deposit with an Umbric
  Dystrochrept.

- In the upper dry valley soils with an argillic horizon
  occur, covered by a colluvial deposit without a cambic
  horizon. (Mollic Hapludalf and Thapto-hapludalfic
  Udorthent).

156

The thickness of the intermediate and young slope deposit is a dominant factor for the classification of the soils. The effectiveness of the erosion on the "intermediate" slope deposit on the upper convex part of the steep lower slopes, after the formation of the argillic horizon seems to be evident because this horizon probably has been removed. These soils should be classified as Umbric Dystrochrepts. Erosion was less effective on the other geomorphic units of the valley slopes, resulting in a thin older "intermediate" slope deposit with a thin argillic horizon on the gentle upper slopes, and a thicker intermediate slope deposit with a relatively thick argillic horizon on the steep lower slopes and dry valley. Also the young slope deposits show the same distribution as the older slope deposit with regard to their thickness: a rather thin layer on the gentle upper slope and the upper convex part of the steep lower slopes and a thicker layer on the steep lower slope. Moreover, a thin young slope deposit was formed in the upper dry valley. On the valley slope in this young slope deposit Dystrochrepts with a low base saturation have developed. In those two places where the young slope deposit was thin, this soil formation also affected (the upper part of) the buried argillic horizon (low base saturation). Consequently the buried part of the soil profile on the gentle upper slope should be classified as Hapludult, while the buried part of the soil profile on the steep lower slope should be classified as Hapludalf.

In the second part of this study the water-rock inter-actions will be discussed. In order to get a complete picture of the chemical weathering process one has to approach the subject from several angles. Consequently the solid phases of the rock and weathering materials and the liquid phase (precipitation-, soil- and springwaters) have been investigated. Due to the mixed character of the weathering materials (slope deposits) no complete weathering sequence could be established by X-ray diffraction- and chemical analyses of the (clay) fraction of the weathering materials. The parent rock consists of quartz- K-mica, albite, chlorite and hematite with minor amounts of rutile, apatite and pyrite. Due to the habitus of the parent rocks, phyllites of quartzphyllites, K-mica and quartz are present in varying amounts. Also rocks with a relatively high CO_2-content (3.5 - 4 %) were found, characterised by important amounts of siderite and a chlorite poor in ferrous iron. On the contrary, the rocks with a very low CO_2 content contain chlorite, which is rich in ferrous iron. Soil profiles on rock types and slope deposits with chlorite rich or poor in ferrous iron have been studied.

From X-ray diffraction analysis it could be established that part of the K-mica was transformed and that vermiculite was formed. Also kaolinite and small amounts of smectite were formed. The chemical analyses of the weathering material shows that chlorite rich in ferrous iron must have been transformed faster than K-mica, while the opposite is true for chlorite poor in ferrous iron. Especially the

minerals rich in ferrous iron like chlorite and siderite
alter considerably by chemical weathering and the liberated
iron is fixed mainly as geothite. Also albite alters by
weathering. The above mentioned mineral transformations
are operating in weathering material *in situ* and in the
slope deposits. In the soil profiles on the narrow
plateau tops and in the young slope deposit with the
Dystrochrept also Al-hydroxy interlayering takes place.
Especially in the soil horizon immediately below the All-Al2
horizon this Al-hydroxy interlayering is a pronounced
feature. This interlayering is probably favoured because
the aluminium (and iron) mobilization by chelating is
limited to the upper soil horizon. At a certain depth
microbiological oxidation of the metal-organic complexes
probably plays an important role, resulting in locally
high Al^{3+} activity and formation of Al-hydroxy interlayered
minerals (Brinkman 1978).

In the Haarts catchment there is no evidence of any
groundwater contribution from outside the drainage basin.
From the water balance it can be concluded that water
leakage is negligible. Therefore it is assumed that the
watershed is watertight. This means that the composition
of the springwaters must be strongly related to the
weathering processes and the residence time of the waters
in the weathering materials and rocks. Although the
chemical composition of the atmospheric precipitation is
strongly influenced by industrial pollution, the contribution of the meteorological input with respect to the
metal cations is limited. Only the chloride, sulphate and
nitrogen input is important, because these constituents
determine the acidity of the precipitation. The driving
force of chemical weathering, the hydrogen ion, in an
industrial area like Western Europe, comes from two sources.
One source is external for the ecosystem (weathering system
+ biomass) and the other internal. The contribution of
the external source to the hydrogen ion input is about 25
per cent if it is assumed that 20 % of the ammonium input
is oxidized to nitrate. The rest of the hydrogen ion contribution is mainly supplied by the mineralization of organic
matter (CO_2-production).

If the upper part of the soil system with the Al-
hydroxy interlayering is excluded, the chemical weathering
reactions seem to be simple dissolution-precipitation
reactions. The rock forming minerals albite,chlorite,
K-mica and siderite are attacked, resulting in an incongruent
dissolution reaction and formation of kaolinite and
geothite and release of sodium, magnesium,(potassium),
aqueous silica and bicarbonate. The composition of the
springwaters indicate that this incongruent dissolution
reaction (kaolinitization stage) seems to be effective
most of the year. However, at low flow conditions (summer
periods) with alkalinities above 1.2 meq/l, partial
equilibria seem to exist between kaolinite and vermiculite
and/or smectite. These partial equilibria have been
evaluated from the curves of the plotted values of the disequilibrium indices (fig. 23, 24). Mass balance
calculations indicate that particularly albite and chlorite

rich in ferrous iron contribute to the release of cations and aqueous silica in the springwaters. Also the dissolution of apatite seems to be important. The precise contribution of K-mica to the release of cations, aqueous silica and weathering products cannot be evaluated accurately, due to the selective uptake of potassium by plant roots and/or adsorption.

The results of the theoretical weathering models compare favourably with the (clay) mineralogy of the saprolites. These models predict a kaolinite, vermiculite and/or smectite and geothite assemblage, especially when the redox potential of the weathering system is governed by the partial oxygen pressure and not by the ferrous/ferric iron couple. Due to the Mg/Al/Si molar ratio in the parent rocks the weathering models predict successively a formation of geothite, kaolinite and a simultaneous precipitation of kaolinite, vermiculite and/or smectite. This sequence has been confirmed by the spring-water chemistry. However, when the theoretical weathering paths are compared with the chemical composition of the soil- and springwater at various hydrological conditions some remarks should be made. These waters indicate a steeper slope of the weathering paths in the kaolinite stability field than those predicted by the theoretical weathering model (fig. 22a, 28, 30). The divergence can be explained by different relative reaction rates of the rock-forming minerals which are unlike those used in the model. Nevertheless if these reaction rates are changed, the predicted succession of formation of weathering products, including the simultaneous formation of kaolinite, vermiculite and/or smectite, can be maintained.

Finally the chemical denudation rate in the Haarts catchment has been calculated. Only the constituents which can be derived from the rocks are taken into account (K, Na, Ca, Mg and SiO_2). If steady state conditions for the biomass are assumed the chemical denudation rate is 62.3 ka/ha.yr. This value indicates that the contribution of the chemical weathering to the output of the dissolved matter which could be derived from rocks was 81.2 percent. The average lowering of rock in the catchment by chemical denudation amounts to approximately 0.5 mm in the last 200 years.

REFERENCES

Adams, W.A., 1976. Experimental evidence of the origin of vermiculite in soils on Lower Palaeozoic sediments. *Soil Sci. Soc. Amer. J.*, 40, 793-796

Alexandre, J., 1957. La restitution des surfaces d'aplanissement tertiarie de l'Ardenne central et ses enseignements. *Ann. Soc. Géo. Belg.*, 81, 333-356

Anderson, V.G., 1945. Some effects of atmospheric evaporation and transpiration on the composition of natural waters in Australia. *J. Proc. Austr. Chem. Inst.*, 12, 41-68 and 83-98

Andriesse, J.P., 1975. *Characteristics and formation of so called Red-Yellow Podzolic soils in the humid tropics (Sarawak-Malaysia)*. K.I.T. Communication 66, Amsterdam, 187 pp

Baeckeroot, G., 1942. *Oesling et Gutland*. Paris

Bakker, J.P., 1948. Over tektogene en morfogene gelijktijdigheid bij de jonge gebergtevorming in West- en Midden-Europa, in het kader van de denudatieve altiplanatie. *Natuurwetenschappelijk Tijdschrift*, 30, 3-53

Barnes, B.S., 1939. The structure of discharge-recession curves. *Trans. Amer. Geophys. Union*, 20, 721-725

Barret, E. & Brodin, G., 1955. The acidity of Scandinavian precipitation. *Tellus*, 7, 251-257

Barton, P.L., Bethke, P.M. & Toulmin, P., 1963. Equilibrium in ore deposits. *Miner. Soc. Amer. Spec. Pap.*, 1, 171-185

Bascomb, C.L., 1968. Distribution of pyrophosphate-extractable iron and organic carbon in soils of various groups. *J. Soil Sci.*, 19, 251-268

Beckwith, R.S. & Reeve, R., 1963. Soluble silica in soils. I. *Aust. J. Soil Res.*, 1, 157-168

Bethune, P. de., 1948. Het Appalachioch relief in Pennsylvanie en in de Ardennen. *Natuurwetenschappelijk Tijdschr.*, 30, 55-64

Bierhuizen, J.F., Abd el Rahman, A.A. & Kuiper, P.J.C., 1960. The effect of nitrogen application and water supply on growth and water requirement of tomato under controlled conditions. *Inst. for Land and Water Management Res., Tech. Bull.*, 16, Wageningen, 12 pp

Bintz, J., 1964. Die Geologie und der variscische Gebirgsbau im Bereich des Pumpspeicherwerkes Vianden. *Publ. Serv. Geol. Luxembourg*, XIV, 77-79

Bintz, J., Muller, A. & Hary, A., 1973. In: *Guide géologique Ardenne Luxembourg*. Masson, Paris, 200 pp

Bischof, W., 1960. Periodical variations of the atmosphere CO_2 content in Scandinavia. *Tellus*, 11, 216-226

Blume, H.P. & Schwertmann, U., 1969. Genetic evaluation of profile distribution of aluminium, iron and manganese oxides. *Soil Sci. Soc. of Amer. Proc.*, 33, 438-444

Bolin, B. (ed)., 1971. *Air pollution across national boundaries. The impact on the environment of sulphur in air and precipitation.* Rept. of the Swedish Prep. Comm. for the U.N. Conf. on human env., Kungl. Boktryckeriet P.A. Norstedt & Sons, Stockholm, 96 pp

Bormann, F.H. & Likens, G.E., 1967. Nutrient cycling. *Science,* 155, 424-429

Boussinesq, J., 1904. Recherches théoriques sur l'écoulement des nappes d'eau infiltrées dans le sol et sur le débit des sources. *J. de Math. Pures et Appliqués,* 10 (5th ser.), 5-78 and 363-394

Breemen, N. van, 1976. *Genesis and solution chemistry of acid sulphate soils in Thailand.* Agric. Res. Rep. 848, Pudoc, Wageningen, 263 pp

Brindley, G.W., 1961. Quantitative analysis of clay mixtures. In: G. Brown (ed), *The x-ray identification and crystal structures of clay minerals.* Mineral Soc., London, 489-514

Brinkman, R., 1977. Surface-water gley soils in Bangladesh: Genesis. *Geoderma,* 17, 111-144

Brinkman, R., 1979. Clay transformations: aspects and equilibrium and kinetics. In: G. H. Bolt (ed), *Soil Chemistry Vo. V B, Physico-chemical models.* Elsevier, Amsterdam, 433-457

Brosset, C., 1976. A method of measuring airborne acidity: its application for the determining of acid content on long-distance transported particles and in drainage water from spruces. *Water, air and soil pollution,* 6, 260-275

Burri, C., 1964. *Petrochemical calculations based on equivalents (methods of Paul Niggli).* Israel Program of Scientific Translations, Jerusalem, 304 pp

Butler Jr., P., 1969. Mineral compositions and equilibria in the metamorphosed iron formation of the Gagnom Region, Quebec. *Canada J. Petrol.,* 10, 56-101

Buurman, P., van der Plas, L. & Slager, S., 1976. A toposequence of alpine soils on calcareous mica schists, Northern Adula Region, Switzerland. *J. of Soil Science,* 27, 395-410

Carroll, D., 1962. *Geochemistry of water: rainwater as a chemical agent of geological processes - a review.* U.S. Geol. Survey Water Supply Paper, 1535 G

Carstea, D.D., Harward, M.E. & Knox, E.G., 1970a. Comparison of iron and aluminium hydroxy interlayers in montmorillonite and vermiculite. I.Formation. *Soil Sci. Amer. Proc.,* 34, 517-521

Carstea, D.D., Harward, M.E. & Knox, E.G., 1970b.
Comparison of iron and aluminium hydroxy interlayers
in montmorillonite and vermiculite. II. Dissolution.
Soil Sci. Amer. Proc., 34, 522-526

Chesselet, R., Morelli, J. & Menard, P.B., 1972. Some
aspects of geochemistry of marine aerosols. In:
*The changing chemistry of the sea. Nobel Symposium
20.* Wiley Interscience, New York, 93-120

Cleaves, T., Godfrey, A.E. & Bricker, O.P., 1970.
Geochemical balance of a small watershed and its
geomorphic implications. *Geol. Soc. Amer. Bull.*,
81, 3015-3032

Coffman, C.B. & Fanning, D.S., 1975. Maryland soils
developed in residuum from chlorite Metabasalt having
high amounts of vermiculite in sand and silt fractions.
Soil Sci. Soc. Amer. Proc., 39, 723-732

Cogbill, C.V. & Likens, G. E., 1974. Acid precipitation
in the northeastern United States. *Water Resources
Res.*, 10, 1133-1137

Dammann, W., 1965. Meteorologische Verdunstungmessungen,
Näherungsformerln und die Verdunstung in Deutschland.
Die Wasserwirtschaft, 55, 315-321

Davies, C.W., 1962. *Ion association.* Butterworths,
London, 190 pp

Deckers, J., 1966. Contribution a l'étude de la
composition et de la capacité de production des sols
de l'Ardenne centrale et de la Famenne orientale.
Pédologie, Mem. 3, 293 pp

Deer, W.A., Howie, R.A. & Zussman, J., 1966. *An introduction to the rock-forming minerals.* Longman, London,
528 pp

Dirven, J.M.C., van Schuylenborgh J & van Breemen, B., 1976.
Weathering of serpentinite in Mantanzas Province,
Cuba. Mass transfer calculations and irreversible
reaction pathways. *Soil Sci. Soc. Amer. Journal*, 40,
901-907

Dov Ashbel, A. Eviatar, Doron, E., Ganor E & Agmon, V.
1965. *Soil temperature in different latitudes and
different climates.* The Hebrew University of Jerusalem,
Jerusalem, 225 pp

Drogue, C., 1972. Analyse statistique des hydrogrammes de
décrues des sources karstiques. *J. Hydrol.*, 15, 49-68

Duchaufour, Ph., 1976. *Atlas écologique des sols du monde.*
Masson, Paris, 178 pp

Duchaufour, Ph., 1977. *Pedologie I: Pédogenèse et
classification.* Masson, Paris, 477 pp

Duchaufour, Ph. & Souchier, B., 1978. Roles of iron and
clay in genesis of acid soils under a humid,
temperate climate. *Geoderma*, 20, 15-26

Dunoyer de Segonzac, G., 1969. *Les mineraux argileux dans la diagenese. Passage au Metamorphisme.* Mem. Serv. Carte Geol. Alsace Lorraine,29, Strasbourg, 320 pp

Dunoyer de Segonzac, G., 1970. The transformation of clay minerals during diagenesis and low-grade metamorphism: a review. *Sedimentology,* 15, 281-346

Edwards, A.M.C., 1975. Dissolved load and tentative solute budgets of some Norfolk catchments. *J. Hydrol.,* 18, 201-217

Ellison, W.D., 1944a. Two devices for measuring soil erosion. *Agric. Eng.,* 25, 53-55

Ellison, W.D., 1944b. Studies of raindrop erosion. *Agric. Eng.,* 25, 181-182

Eriksson, E., 1952a. Composition of atmospheric precipitation. I. Nitrogen compounds. *Tellus,* 4, 213-232

Eriksson, E., 1952b. Composition of atmospheric precipitation. II. Sulphur, chloride, iron compounds. Bibliography. *Tellus,* 4, 280-303

Eriksson, E., 1955. Airborne salts and the chemical composition of river waters. *Tellus,* 7, 243-250

Eriksson, E., 1959. The yearly circulation of chloride and sulphur in nature, Part I. Meteorological, geochemical and pedological implications. *Tellus,* 11, 375-403

Eriksson, E., o960. The yearly circulation of chloride and sulphur in nature, Part II. Meteorological, geochemical and pedological implications. *Tellus,* 12, 63-109

Eriksson, E., 1963. The yearly circulation of sulphur in nature. *J. Geophys. Res.,* 68, 4001-4008

Eriksson, E., 1969. *Sulphur dioxide and the acidification of precipitation - facts and speculations* (in Swedish). I.V.L. - Publ. A 28, Stockholm

Faber, R., 1971. *Climatologie du Grand-Duche de Luxembourg.* Publication du Musee d'Histoire Naturelle et de la Soc. des Naturalistes Luxembourgeois, Luxembourg, 48 pp

Faber, R. & Schmidt, E., 1963. Excursion du dimanche, 27 Octobre 1963. *Soc. de Nat. Luxembourgeois Bull.,* 68, 220-224

F.A.O., 1968. *Guidelines for soil profile description.* Soil Survey & Fertility Branch, Land & Water Development Division, F.A.O., Rome, 53 pp

F.A.O., 1974. F.A.O.-UNESCO, *Soil Map of the World. Vol. 1, Legend.* UNESCO, Paris, 59 pp

Ferraris, Le Comte de,1777, reissued 1965-1970, *Carte de Cabinet des Pays-Bas Autrichiens.* Bibl. Royale de Belgique, Bruxelles

Feth, J.H., Robertson, C.E. & Polzer, W.L., 1964. *Sources of mineral constituents in water from granitic rocks - Sierra Nevada, California and Nevada.* Geochemistry of water: U.S. Geol. Survey Water Supply Paper 1535, 70 pp

Firbas, F., 1949. *Spät- und nacheiszeitliche Waldgeschichte Mitteleuropas nördlich der Alpen. Bd. I. Allgemeine Waldgeschichte.* Verlag Gustav Fischer, Jena, 480 pp

Firbas, F., 1952. *Spät- und nacheiszeitliche Waldgeschichte Mitteleuropas nördlich der Alpen, Bd. II. Waldgeschichte der einzelnen Landschaften.* Verlag Gustav Fischer, Jena, 265 pp

Fisher, D.W., Gambell, A.W., Likens, G.E. & Bormann, F.H., 1968. Atmospheric contributions to water quality of streams in the Hubbard Brook Experimental Forest, New Hampshire. *Water Resources Res.*, 4, 1115-1126

Foster, M.D., 1960. *Interpretation of the composition of trioctahedral micas.* U.S. Geol. Surv. Prof. Paper, 354-B, 11 pp

Foster, M.D., 1962. *Interpretation of the composition and classification of chlorites.* U.S. Geol. Surv. Prof. Paper, 414-A, 27 pp

Foster, M.D., 1963. Interpretation of the composition of vermiculites and hydrobiotites. *Clays and Clay Miner. Proc.*, 10, 70-89

Forkaziewicz, J. & Paloc, H., 1965. Le régime de Tarissement de la France de la vis d'étude préliminaire. *IASM, Publ.* 74, 213-226

Fourmarier, P. (ed)., 1954. *Prodome d'une description géologique de la Belgique.* Soc. Géol. Belgique, Liège, 826 pp

Friend, J.P., 1973. The global sulphur cycle. In: *Chemistry of the lower atmosphere.* Plenum Press, New Yori, 177-201

Fritz, B., 1975. *Etude thermodynamique et simulations des réactions entre minéraux et solutions applications à la geochimie des alterations et des eaux continentales.* Mem. Sciences Geologiques 41, Strassbourg, 152 pp

Galloway, J.N. & Likens, G.E., 1976. Calibration of collection procedures for the determination of precipitation chemistry. *Water, air and soil pollution,* 6, 241-258

Galloway, J.N. & Likens, G.E., 1978. The collection of precipitation for chemical analysis. *Tellus,* 30, 71-82

Galloway, J.N., Likens, G.E., & Edgerton, E.S., 1976a. Acid precipitation in the northeastern United States: pH and acidity. *Science,* 194, 722-724

Galloway, J.N., Likens, G.E. & Edgerton, E.S., 1976b. Hydrogen ion specification in the acid precipitations of the northeastern United States. *Water, air and soil pollution,* 6, 423-433

Gambell, A.W. & Fisher, D.W., 1966. *Chemical composition of rainfall in north-eastern Carolina and southwestern Virginia.* U.S. Geol. Survey Water Supply Paper, 1535 K, 41 pp

Gamble, E.E. & Daniels, R.B., 1972. Iron and silica in water, acid ammonium oxalate and dithionite extracts of some North Carolina Coastal Plain Soils. *Soil Sci. Soc. Amer. Proc.,* 36, 939-943

Gardner, W.R. & Ehlig, C.F., 1963. The influence of soil water on transpiration by plants. *J. Geophys. Res.,* 68, 5719-5724

Garrels, R.M. & Mackenzie, F.T., 1967. Origin of the chemical compositions of some springs and lakes. In: R. G. Gould (ed), *Equilibrium concepts in natural water systems.* Adv. in Chem. Ser. 67, Amer. Chem. Soc., Washington DC, 222-242

Gorham, E., 1955. On the acidity and salinity of rain. *Geochim, et Cosmochim. Acta,* 7, 231-239

Gorham, E., 1958. The influence and importance of daily weather conditions in the supply of chloride, sulphate and other ions to fresh waters from atmospheric precipitation. *Phil. Trans. Roy. Soc. London,* Ser. B, 241, 147-178

Gorham, E., 1961. Factors influencing the supply of major ions to inland waters, with special reference to the atmosphere. *Geol. Soc. Amer. Bull.,* 72, 795-840

Gorham, E., 1976. Acid precipitation and its influence upon aquatic ecosystems. An overview. *Air, water and soil pollution,* 6, 457-481

Granat, L., 1972a. On the relation between pH and the chemical composition in atmospheric precipitation. *Tellus,* 24, 550-560

Granat, L., 1972b. *Deposition of sulphate and acid with precipitation over northern Europe.* Report A.C. -20, Inst. of Meteorology, University of Stockholm, 18 pp

Guillien, Y., 1951. Les grèzes litées de Charente. *Rev. Géogr. Pyrénées et S.O.,* 22

Guillien, Y., 1964. Grèzes litées et bancs de niege. *Geol. en Mijnb.,* 43, 103-112

Gullentops, F., 1954. Contributions à la chronologie du pleistocène et des formes du relief en Belgique. *Mem. Inst. Géol. Univ. de Louvain,* 18

Hall, F.R., 1968. Base-flow recessions - a review. *Water Resources Research,* 4, 973-983

Hashimoto, I. & Jackson, M.L., 1960. Rapid dissolution of allophane and kaolinite-halloysite after dehydration. *Clays and Clay Miner.,* 7, 102-103

Haude, W., 1954. Zur praktischen Bestimmung der aktuellen und potentiellen Evaporation und Evapotranspiration. *Mitt. Deutsch. Wetterdienst,* 8, 3-22

Haude, W., 1963. Bestimmung der Verdunstung und des Wasserhaushaltes in Trockengebieten des Vorderen Orients zwischen Nil und Euphrat. *Die Wasserwirtschaft,* 53, 427-438

Heath, R.C. & Trainer, F.W., 1968. *Introduction to groundwater hydrology.* Wiley & Sons Inc., New York, 284 pp

Helgeson, H.C., 1968. Evaluation of irreversible reactions in geochemical processes involving minerals and aqueous solutions. I. Thermodynamic relations. *Geochimica et Cosmochimica Acta,* 32, 853-877

Helgeson, H.C., 1969. Thermodynamics of hydrothermal systems at elevated temperatures and pressures. *Am. J. Science,* 267, 729-804

Helgeson, H.C., 1971. Kinetics of mass transfer among silicates and aqueous solutions. *Geochimica et Cosmochimica Acta,* 35, 421-469

Helgeson, H.C., Garrels, R.M. & Mackenzie, F.T., 1969. Evaluation of irreversible reactions in geochemical processes involving minerals and aqueous solutions. II. Applications. *Geochimica et Cosmochimica Acta,* 33, 455-481

Hermans, W.F., 1955. Description et genese des depots meubles de surface et du relief de l'Oesling. Serv. Geol. de Luxembourg, Luxembourg, 94 pp

Hewlett, J.D. & Hibbert, A.R., 1967. Factors affecting the response of small watersheds to precipitation in humid areas. In: W. E. Sopper & H. W. Lull (eds)., *Forest Hydrology,* Pergamon Press, New York, 275-290

Higashi, T. & H. Ikeda, 1973. Dissolution of allophane by acid oxalate solution. *Clay Sci.,* 4, 205-212

Holmgren, G.G.S., 1967. A rapid citrate-dithionite extractable iron procedure. *Soil Sci. Soc. Amer. Proc.,* 31, 210-211

Hornbeck, J.W., Likens, G.E. & Eaton, J.S., 1977. Seasonal patterns in acidity of precipitation and their implications for forest stream ecosystems. *Water, air and soil pollution,* 7, 355-365

Horton, R.E., 1945. Erosional development of streams and their drainage basins: hydrophysical approach to quantitative morphology. *Bull. Geol. Soc. Amer.,* 56, 275-370

Hudson, N., 1971. *Soil conservation.* Batsford Ltd., London, 320 pp

Imeson, A.C., 1976. Some effects of burrowing animals on slope processes in the Luxembourg Ardennes. Part 1. The excavation of animal mounds in experimental plots. *Geografiska Annaler,* 58, Ser. A., 115-125

Imeson, A.C., 1977. Splash erosion, animal activity and sediment supply in a small forested Luxembourg catchment. *Earth Surface Processes*, 2, 153-160

Imeson, A.C. & Jungerius, P.D., 1974. Landscape stability in the Luxembourg Ardennes as exemplified by hydrological (micro)- pedological investigations of a catena in an experimental watershed. *Catena*, 1, 273-295

Imeson, A.C. & Kwaad, F.J.M.P., 1976. Some effects of burrowing animals on slope processes in the Luxembourg Ardennes. Part 2. The erosion of animal mounds by splash under forest. *Geografiska Annaler*, 58, Ser. A, 317-328

Imeson, A.C. & Verstraten, J.M., 1979. Runoff and solute characteristics of a forested watershed in the Luxembourg Ardennes. *In preparation*.

Jackson, M.L., 1964. Chemical compositions of soils. In: F. Bear (ed)., *Chemistry of soil*. Van Nostrend Rheinhold Comp., New York, 65-77

Jackson, M.L. & Sherman, G.D., 1953. Chemical weathering of minerals in soils. *Advances in Agronomy*, V, 219-318

Juans, F.H.T. & Johnsson, N.M., 1967. Cycling of chloride through a forested watershed in New England. *J. Geographical Res.*, 72, 5641-5647

Junge, C.E., 1958. Atmospheric chemistry. *Advances in geophysics*, 4, 1-108, Reinhold Publ. Corp, New York

Junge, C.E., 1963. *Air chemistry and radioactivity*. Academic Press, New Yori, 382 pp

Junge, C.E. & Werby, R.T., 1958. The concentration of chloride, sodium, potassium, calcium and sulphate in rainwater over the United States. *J. Meteorology*, 15, 417-425

Kellogg, W.W., Cattle, R.D., Allen, E.W., Lazlus, A.L. & Martell, E.A., 1972. The sulphur cycle. *Science*, 175, 587-595

Kessler, J. & Oosterbaan, R., 1964. Determining hydraulic conductivity of soils. In: *Drainage principles and applications*, Publ. 16, III, I.I.L.C., Wageningen, The Netherlands, 253-296

Kittrick, J.A., 1971. Stability of montmorillonites. I. Belle Fourche and Clay Spur Montmorillonite. *Soil Sci. Soc. America Proc.*, 35, 140-145

Kittrick, J.A., 1973. Mica-derived vermiculites as unstable intermediates. *Clays and Clay Miner.*, 21, 479-488

Konrad, H.J. & Wachsmut, W., 1973. Zur Lithologie und Tektonik des Unterdevons im südlichen Oesling Luxemburgs. *Publ. Serv. Geol. Luxembourg*, Bull. 5, 1-20

Korzhinskii, D.S., 1959. *Physicochemical basis of the analysis of the paragenesis of minerals*. Consultants Bureau Inc., New York, 142 pp

Korzhinskii, D.S., 1963. Thermodynamic potentials of open systems whose acidity and reduction potentials are determined by external conditions. *Dokl. Acad. Sci. USSR, AGI Translation,* 152, 175-177

Korzhinskii, D.S., 1964. An outline of metasomatic processes. *Int. Geol. Rev.,* 6, 1713-1734; 1920-1952; 2169-2198

Korzhinskii, D.S., 1965. The theory of systems with perfectly mobile components and processes of mineral formation. *American J. Science,* 263, 193-205

Kramer, P.J., 1952. Plant and soil water relations on the watershed. *J. Forestry,* 50

Krayenhoff van de Leur, D.A., 1973. Rainfall-runoff relations and computational models. In: *Drainage principles and applications* II. Theories of field drainage and watershed runoff. I.L.R.I. Publ. 16, II. Wageningen, 245-320

Kubler, B., 1964. Les argiles, indicateurs de métamorphisme. *Res. Inst. Franc. Pétrole,* 19, 1093-1112

Kubler, B., 1966. La cristallinité de l'illite et les zones tout à fait supérieures du métamorphisme. In: *Colloque sur les Etages tectoniques.* A la Baconnière, Nauchâtel, 105-122

Kubler, B., 1967. Anchimétamorphisme et schistosité. *Bull. Centre Rech.,* Pau-S.N.P.A., 1, 259-278

Kwaad, F.J.P.M., 1977. Measurements of rainsplash erosion and the formation of colluvium beneath deciduous woodland in the Luxembourg Ardennes. *Earth Surface Processes,* 2, 161-173

Kwaad, F.J.P.M. & Mücher, H.J., 1976. First Benelux Colloquium on geomorphological processes. *Excursion Guide,* Louvain, 25-28

Kwaad, F.J.P.M., & Mücher, H.J., 1977. The evolution of soils and slope deposits in the Luxembourg Ardennes near Wiltz. *Geoderma,* 17, 1-37

Kwaad, F.J.P.M., & Mücher, H.J., 1978. Colluvium on arable land near Berlé, northern Luxembourg. *Geoderma, in press*

Lahr, E., 1964. *Temps et climat.* Serv. Météorol. et Hydrol. Natural de Luxembourg, Luxembourg, 290 pp

Langmuir, D., 1971. Particle sixe effect on the reaction geothite→hematite + water. *American J. Science,* 271, 147-156

Langmuir, D., 1972. Correction. Particle size effect on the reaction geothite→hematite + water. *American J. Science,* 272, 972

Langmuir, D. & Whittemore, D.O., 1971. Variations in the stability of precipitated ferric oxyhydroxides. In: J. D. Hem (ed), *Non-equilibrium systems in natural water chemistry.* Adv. Chem. Soc., 106, Am. Chem. Soc., Washington DC, 209-234

Lefevre, M.A., 1938. Surfaces d'asplanissement de l'Ardenne belge et de son avant-pays. *Bull. Soc. Belge Etudes Geogr.*, XXI, 41-68

Likens, G.E., Bormann, F.H. & Johnson, N.M., 1972. Acid rain. *Environment*, 14, 33-40

Likens, G.E. & Bormann, F.H., 1974. Acid rain: a serious regional environmental problem. *Science*, 184, 1176-1179

Likens, G.E., Bormann, F.H., Eaton, J.S., Pierce, R.S. & Johnson, N.M., 1976. Hydrogen ion input to the Hubbard Brook Experimental Forest, New Hampshire, during the last decade. *Water, soil and air pollution*, 6, 435-445

Likens, G.E., Bormann, F.H., Pierce, R.S., Eaton, J.S. & Johnson, N.M., 1977. *Biogeochemistry of a forested ecosystem.* Springer Verlag, New York, 146 pp

Linsley, R.K., Kohler, M.A. & Paulhaus, J.L.H., 1958. *Hydrology for Engineers.* McGraw Hill, New York, 340 pp

Lucius, M., 1940. Der Werdegang des Luxemburger Sedimentationsraumes seit dem Ausgang des Paläozoikums *Livre jubilaire du cinquantenaire de la Soc. des Nat. Lux.*, facs. 1, Luxembourg

Lucius, M., 1950a. Erläuterungen zur geologischen Karte von Luxemburg: *Das Oesling.* Veröffentl. Lux. Geol. D., Bd. VI, Luxembourg, 174 pp

Lucius, M., 1950b. La notion de péneplaine et le modele du terrain de l'Ardenne Luxembourgeaoise (Oesling). *Bull. Soc. Nat. Lux.*, N.S., 44, 108-122

Lucius, M. & Bintz, J., 1960. Amenagement hydroelectrique de l'Our. *Rev. Techn. Luxembourgeoise Oct-Dec.*

Lyshede, J.M., 1955. *Hydrologic studies of Danish watercourses.* Folia Geographica Danica, 6, 155 pp

Macar, P., 1954. L'evolution geomorphologique de l'Ardenne. *Bull. Soc. Roy. Belge Geogr.*, t, 78

Machta, L., 1972. The role of the oceans and biosphere in the carbon dioxide cycle. In: D. Dryssen & D. Jagner (eds), *The changing chemistry of the oceans.* Nobel Symposium 20, Wiley Interscience, New York

Mackenzie, F.T. & Gees, R., 1971. Quartz: synthesis at earth-surface conditions. *Science*, 173, 533-535

Makkink, G.F. & Van Heemst, H.D.J., 1956. The actual evapotranspiration as a function of the potential evapotranspiration and the soil moisture tension. *Neth. J. Agric. Sci.*, 4, 67-72

McKeague, J.A. & Cline, M.G., 1963a. Silica in soils I. The form and concentration of dissolved silica in aqueous extracts of some soils. *Can. J. Soil Science*, 43, 70-82

McKenzie, R.M., 1977. Manganese oxides and hydroxides. In: J. B. Dixon & S. B. Weed (eds): Minerals in soil environments. *Soil Sci. Soc. of America,* Madison, 181-194

Mehra, O.P. & Jackson, M.L., 1960. Iron oxide removal from soils and clays by a dithionite-citrate system with sodium bicarbonate buffer. *Clays and Clay Miner.,* 7, 317-327

Melillo, J.M., 1977. *Nitrogen dynamics in an aggrading northern hardwood forest ecosystem.* Ph.D thesis, Yale University, New Haven, Connecticut

Meszaros, E., 1966. On the origin and composition of atmospheric calcium compounds. *Tellus,* 18, 262-265

Millot, G., 1964. *Géologie des argiles.* Masson & Cie, Paris, 499 pp

Mohr, E.C.J, van Baren, F.A. & can Schuylenborgh, J., 1972. *Tropical soils.* Mouton-Ichtiar Baru - Van Hoeve, The Hague, 481 pp

Mückenhausen, E. & Mitarbeiter, 1976. *Entstehung, Eigenschaften und Systematik der Böden der Bundesrepublik Duetschland.* Deutsch Landw. Ges., Frankfurt

Muller, R.F. & Saxene, S.K., 1977. *Chemical petrology.* Springer Verlag, New York, 394 pp

Nepper, D.M., 1904. *Die landwirtschaftliche Benutzung des Grund und Bodens in dem Grossherzogtum Luxemburg.* Bonn

Nutbrown, D.A. & Downing, R.A., 1976. Normal-mode analysis of the structure of baseflow recession curves. *J. Hydrology,* 30, 327-340

Oden, S., 1976. The acidity problem - an outline of concepts. *Water, soil and air pollution,* 6, 137-166

Ormsby, W.C. & Sand, L.B., 1954. *An analytical look for mixed-layer aggregates.* Proc. Nat. Conf. Clays Clay Miner., 2nd Nat. Acad. Sci. Nat. Res. Course Publ., 327, 254-263

Paces, T., 1973. Actual mineral surfaces: origin and possible effect on trace elements in natural water systems. In: D. D. Hemphil (ed), *Trace substances in environmental health,* VI, 361-368

Pa Ho Hsu, 1977. Aluminium hydroxides and oxyhydroxides. In: J. B. Dizon & S. B. Weed (eds), *Minerals in the soil environment.* Soil Sci. Soc. America, Madison, 99-144

Parizek, R.R. & Lane, B.E., 1970. Soil water sampling using pan and deep pressure vacuum lysimeters. *J. of Hydrology,* 11, 1-22

Peck, A.J. & Hurle, D.H., 1973. Chloride balance of some farmed and forested catchments in south-western Australia. *Water Resources Res.,* 9, 648-657

Penman, H.L., 1949. The dependence of transpiration on weather and soil conditions. *J. Soil Science*, 1, 74-89

Penman, H.L., 1965. Evaporation: an introductory survey. *Neth. J. Agric. Sci.*, 42, 286-292

Penman, H.L., 1969. The role of vegetation in soil water problems. In: P. E. Rytema & H. Wassink (eds), *Water in the unsaturated zone*, IASH-UNESCO, Paris, 49-69

Piket, J.J.C., 1960. *Het Oesling landschap rondom Hosingen*. PhD thesis, Utrecht, 145 pp

Pissart, A., 1961. Les aplanissements tertiaires et les surfaces d'érosion anciennes de l'Ardenne du sud-ouest. *Ann. Soc. Geol. Belg.*, 85

Plas, L. van der & van Schuylenborgh, J., 1970. Petro-chemical calculations applied to soils (with special reference to soil formation). *Geoderma*, 4, 357-385

Reiche, P., 1950. *A survey of weathering processes and products*. Univ. New Mexico Publication in Geology, 3. Rv. Ed.

Rich, C.I., 1968. Hydroxy-interlayers in expansible layer silicates. *Clays and Clay Miner.*, 16, 15-30

Ridder, T.B., 1978. *Over de chemie van de neerslag*. K.N.M.I., W.R. 78-4, De Bilt, 45 pp

Riezebos, P.A. & Slotboom, R.T., 1974. Palynology in the study of present-day hillslope development. *Geol. en Mignbouw*, 53, 436-448

Riezebos, P.A. & Slotboom, R.T., 1978. Pollen analysis of the Husterbaach peat (Luxembourg): its significance for the study of subrecent geomorphological events. *Boreas*, 7, 75-82

Riley, J.P. & Skirrow, G., 1975. *Chemical oceanography*, vol. 1., 2nd edition, Academic Press, London, 606 pp

Robie, R.A. & Waldbaum, D.R., 1968. *Thermodynamic properties of minerals and related substances at $298.15°K$ $(25.5°)$ and one atmosphere (1.013 bars) pressure and higher temperatures.* U.S. Geol. Survey Bull. 1259, Washington DC, 256 pp

Robie, R.A., Hemingway, B.S. & Fisher, J.R., 1978. *Thermodynamic properties of minerals and related substances at $298.15°K$ and 1 bar (10^5 Pascals) pressure and at higher temperatures.* U.S. Geol. Surv. Bull. 1452, Washington DC, 456 pp

Robinson, E. & Robbins, R.C., 1970. Gaseous nitrogen compound pollutants from urban and natural sources. *J. Air Pollution Control Ass.*, 20, 303-306

Ross, G.J., 1968. Structural decomposition of an ortho-chlorite during its acid dissolution. *Can. Miner.*, 9, 522-530

Ross, G.J., 1969. Acid dissolution of chlorites: release of magnesium, iron and aluminium and mode of acid attack. *Clays and Clay Miner.*, 17, 347-354

Ross, G.J., 1975. Experimental alteration of chlorites into vermiculites by chemical oxidation. *Nature,* 255, 133-134

Ross, G.J. & Kodama, H., 1974. Experimental transformation of a chlorite into a vermiculite. *Clays and Clay Miner.*, 22, 205-211

Scheffer, F. & Schachtschabel, P., 1976. *Lehrbuch der Bodenkunde.* 9th edition, Enke Verlag, Stuttgart, 394 pp

Schnitzer, M. & Hansen, E.H., 1970. Organo-metallic interactions in soils, 8. An evaluation of methods for the determination of stability constants of metal-fulvic acid complexes. *Soil Science,* 109, 333-340

Schnitzer, M. & Khan, S.U., 1972. *Humic substances in the environment.* Marcel Dekker Inc., New Yori, 327 pp

Schnitzer, M. & Skinner, S.I.M., 1963. Organo-metallic interactions in soils. 1. Reactions between a number of metal ions and the organic matter of a podzol B h horizon. *Soil Science,* 96, 86-93

Schnitzer, M. & Skinner, S.I.M., 1964. Organo-metallic interactions in soils. 3. Properties of iron- and aluminium-organic matter complexes, prepared in the laboratory and extracted from a soil. *Soil Science,* 98, 197-203

Schoeller, M., 1961a. Calcul du bilan des nappes d'eau des "Sables des Landes" en utilisant la teneur en chlore de l'eau des nappes et celle de l'eau de pluie. *C.R. Acad. Sci. de Paris,* t. 253, 1598-1599

Schoeller, M., 1961b. Calculs de l'évapotranspiration par la methode de Thornthwaite et par celle des chlorures Concordance des résultants. *C.R. Acad. Sci. de Paris,* t. 253, 3014-3015

Schoeller, M., 1963a. Les lois de la concentration en chlore des eaux souterraines par evaporation et par evapotranspiration. *C.R. Acad. Sci. de Paris,* t. 256, 3024-3025

Schoeller, M., 1963b. Les lois de la concentration en chlore des eaux souterraines par dissolution dans un terrain non lessive en amont. *C.R. Acad. Sci. de Paris,* t, 256, 3195-3196

Schwertmann, U., 1964. Differenzierung der Eisenoxide des Bodens durch Extraction mit Ammoniumoxalat-Lösung. *Z. f. Pflanzenernähr., Dung., Bodenkunde,* 105, 194-202

Schwertmann, U., 1966. Inhibitory effect of soil organic matter on the crystallinization of amorphous ferric hydroxide. *Nature,* 212, 645-646

Schwertmann, U., 1969. Der Einfluss einfacher organischer Anionen auf die Bildung von Goethit und Hämatit aus amorphen Fe(III) - Hydroxid. *Geoderma*, 3, 207-214

Schwertmann, U., Fischer, W.R. & Papendorf, H., 1968. The influence of organic compounds on the formation of iron oxides. *Int. Congr. Soil Science Transactions* 9th, (Adelaide, Austr.), 1, 645-655

Schwertmann, U. & Taylor, R.M., 1977. Iron oxides. In: J. B. Dixon & S. B. Weed (eds), *Minerals in soil environments*. Soil Science Society of America, Madison, 145-180

Sevink, J., 1974. *Landscape evolution and soils of the Southwestern Velay* (France). Publ. Fys. Geogr. Bodemk. Lab, 21, Amsterdam, 183 pp

Siegel, S., 1956. *Non-parametric statistics for the behavioral sciences*. McGraw-Hill Book Comp. Inc., New York

Singh, K.P. & Stall, J.B., 1971. Derivation of base flow recession curves and parameters. *Water Resources Res.*, 7, 292-303, 312 pp

Söderlund, R. & Svensson, B.H., 1976. The global nitrogen cycle. In; B. H. Svensson & R. Söderlund (eds), *Nitrogen, Phosphorus and Sulphur - Global Cycles*. SCOPE Report 7, Ecological Bulletins no. 222, Stockholm, 192 pp

Soil Survey Staff, 1975. *Soil Taxonomy - a basic system of soil classification for making and interpreting soil survey*. Agricultural Handbook no. 436, S.C.S., U.S.D.A., Washington, 754 pp

Steffes, M & G., 1965. *La sidérurgie Luxembourgeoise de l'époque anterieure à 1840*. Luxembourg, Bourg-Bourger

Stevens, Ch., 1959. La geomorphologie de St. Hubert. *Bull. Soc. Geol. Belge*, 18, 152 pp

Stevenson, J.F., Dharival, A.P. & Chondhri, M.B., 1958. Further evidence for naturally occurring fixed ammonia in soils. *Soil Science*, 85, 42-46

Stumm, W. & Morgan, J.J., 1970. *Aquatic chemistry*. Wiley, New York, 583 pp

Thompson, J.B. Jr., 1955. The thermodynamic basis for the mineral facies concept. *American J. Science*, 253, 65-103

Thompson, J.B. Jr., 1959. Local equilibrium in metasomatic processes. In: P. H. Abelson (ed.), *Researches in geochemistry*, vol. 1, Wiley, New York, 427-457

Thornthwaite, C.W., 1954. A re-examination of the concept and measurements of potential evapotranspiration. *Publ. in Climat.*, 7(1), 200-209

Tokashiki, Y. & Wada, K., 1975. Weathering implications of the mineralogy of clay fractions of two Andosols, Kyushu. *Geoderma*, 14, 47-62

Turner, F.J., 1968. *Metamorphic petrology - mineralogical and field aspects*. McGraw-Hill, New York, 403 pp

Uhlig, S., 1954. Zur Bestimmung der potentiellen Verdunstung bewachsener Bodens. *Die Wasserwirtschaft*, 44, 309-315

Veen, A.W.L., 1970. *On geogenesis and pedogenesis in the old coastal plain of Surinam* (South America). Publ. Fys. Geogr. Bodemk. Lab., 14, Amsterdam, 176 pp

Velde, B., 1977. *Clays and clay minerals in natural and synthetic systems*. Developments in Sedimentology, 21, Elsevier, Amsterdam, 218 pp

Verhoef, P., 1966. *Geomorphological and pedological investigations in the Redange-sur-Attert area (Grand-Duchy de Luxembourg)*. Publ. Fys. Geogr. Bodemk. Lab., 8, Amsterdam, 531 pp

Vermeulen, A.J., 1977. *Immissie-onderzoek met behulp van regenvangers: opzet, ervaringen en resultaten*. Prov. Waterstaat van Noord-Holland, Haarlem, 109 pp

Verstraten, J.M., 1977. Chemical erosion in a forested watershed in the Oesling, Luxembourg. *Earth Surface Processes*, 2, 175-184

Verstraten, J.M., 1979a. Theoretical mass transfer calculation models for low-grade metamorphic rocks. *In preparation*

Verstraten, J.M., 1979b. Gibbs free energy of formation of two vermiculites calculated from dissolution experiments. *In preparation*

Viehmeyer, F.J., 1927. Some factors affecting the irrigation requirements of deciduous orchards. *Hilgardia*, 2, 125-184

Viehmeyer, F.J. & Hendrikson, A.H., 1927. Soil moisture conditions in relation to plant growth. *Plant Physiol.*, 2, 71-82

Viehmeyer, F.J. & Hendrikson, A.H., 1955. Does transpiration decrease as the soil moisture decreases? *Trans. American Geophys. Union*, 36, 426-428

Visser, W.C., 1963. Soil moisture content and evapotranspiration. *Inst. for Land and Water Management Res., Technical Bulletin*, 31, Wageningen

Visser, W.C., 1964. Moisture requirements of crops and rate of moisture depletion of the soil. *Inst. for Land and Water Management Res., Technical Bulletin*, 32, Wageningen, 21 pp

Wada, K., 1977. Allophane and imogolite. In: J. B. Dixon & S. B. Weed (eds): *Minerals in soil environments*, Soil Science Soc. of America, Madison, 603-638

Wada, K. & Greenland, D.J., 1970. Selective dissolution and differential infrared spectroscopy for characterization of "amorphous" constituents in soil clays. *Clay Miner.*, 8, 241-254

Wada, K. & Higashi, T., 1976. The categories of aluminium and iron-humus complexes in andosoils determines by selective dissolution. *J. Soil Science,* 27, 357-368

Wada, K. & Wada, S., 1976. Clay mineralogy of the B-horizon of two Hydrandepts, a Torrox and a Humit-ropept in Hawaii. *Geoderma,* 16, 139-157

Walker, G.F., 1975. Vermiculites. In: J. E. Gieseking (ed), *Soil components,* volume 2, Springer Verlag, Berlin, 155-190

Ward, R.C., 1975. *Principles of hydrology.* McGraw-Hill, London, 367 pp

Wartena, L. & Veldman, E.C., 1961. Estimation of basic irrigation requirements. *Neth. J. Agric. Sci.,* 9, 293-298

Waugh, J.R., 1970. Base-flow recessions as an index of representativeness in the hydrological regions of Northland, New Zealand. *IASH-UNESCO Symposium on the results of research in representativeness and experimental basins.* Wellington, New Zealand, 603-613

Weaver, C.E., 1956. The distribution and identification of mixed-layer clays in sedimentary rocks. *Amer. Mineral.,* 41, 202-221

Weaver, C.E., 1958. A discussion on the origin of clay minerals in sedimentary rocks. *Clays and Clay Minerals* (5th Nat. Conf. 1956), 159-173

Weaver, C.E., 1960. Possible uses of clay minerals in the search for oil. *Clays and Clay Minerals* (8th Nat. Conf. 1958), 214-227

Weaver, C.E., 1961. Clay mineralogy of the Late Cretaceous rocks of the Washaki Basin. *Wyoming Geological Association Guidebook,* 16, 145-152

Weaver, C.E. & Beck, K.C., 1971. *Clay water diagenesis during burials: how mud becomes gneiss.* Geological Society of America Special Paper 134, Washington DC, 96 pp

Weaver, C.E. & Pollard, L.D., 1975. *The chemistry of clay minerals.* Developments in sedimentology 15, Elsevier, Amsterdam, 213 pp

Weber, K., 1972. Kristallinität des Illits in Tonschiefern und andere Kriterien schwacher Metamorphose im nordöstlichen Rheinischen Schiefergebirge. *N. Jb. Geol. Paläont. Abh.,* 141

Weyman, D.R., 1974. Runoff processes, contributing area and streamflow in a small upland catchment. *Institute British Geographers Special Publication* 6, 33-43

Whipkey, R.Z., 1965. Subsurface stormflow from forested slopes. *Bull. Int. Assoc. Sci. Hydrol.,* 10, 74-85

Whitehead, H.C. & Feth, J.H., 1964. Chemical composition of rain, dry fallout and bulk precipitation at Melmo Park, California 1957-1959. *J. Geophys. Research,* 69, 3319-3333

Winkler, H.G.F., 1974. *Petrogenesis of metamorphic rocks.* Springer Verlag, New York, 320 pp

Wood, W.W., 1973. A technique using porous cups for water sampling at any depths in the unsaturated zone. *Water Resources Res.,* 9, 486-488

Zeeuw, J.W., 1973. Hydrograph analysis for areas with mainly groundwater runoff. In: *Drainage principles applications* II. Theories of field drainage and watershed runoff. I.L.R.I. Publication 16(2), Wageningen, 321-357

Zen, E-an, 1963. Components, phases and criteria of chemical equilibria in rocks. *American J. Science,* 261, 929-942

Zen, E-an, 1966. *Construction of pressure-temperature diagrams for multicomponent systems after the method of Schreinemakers - a geometric approach.* U.S. Geol. Bull. 1225

Zen, E-an, 1972. Gibbs free energy, enthalpy, and entropy of ten rock-forming minerals. Calculations, discrepancies, implications. *American Mineral.,* 57, 524-553

APPENDICES

APPENDIX 1

PROCEDURE AND METHODS
FOR SOIL AND WATER SAMPLES

1. Field investigations and sampling procedures

1.1 *Descriptions*

All soil profiles were described from freshly dug profile
pits, using "Guidelines for soil profile descriptions"
issued by FAO, Rome and Munsell Soil Color Charts (Munsell
Color Co., Baltimore, USA). Brief descriptions were made of
profiles using samples obtained with an "Edelman-type" auger.

1.2 *Soil samples*

Of each soil horizon representative large bulk samples
were assembled from sub-samples taken randomly from each
distinguished horizon. These samples were dried at $50^{\circ}C$
and after slightly quatering the fraction > 2 mm was
determined. Grain size distribution, pH, organic carbon,
exchangeable cations, dithionite extractable Fe, Al, Mn and
Si, oxalate extractable Fe, Al, Mn and Si, pyrophosphate
extractable Fe, Al and Mn were estimated for the fine earth
fraction after thorough homogenization (quatering and
splitting). Also an estimation was carried out for the
elemental analyses of the fine earth and clay fraction and
for the mineralogy of the clay fraction.

1.3 *Water samples*

Spring and riverwaters were collected directly in
500 cm^3 polyethlyene bottles. Soil water samples were
obtained with the help of porous cup soil water samplers
(Wood 1973) and with PVC tubes which were perforated on
the upper side and served as drains. With the porous cup
soil water sampler water was collected with a tension of
less than 0.8 bar (pF < 2.9). Alkalinity, pH and
electrical conductivity were measured immediately in the
field. Water samples were filtered by pressure filtration
through a 0.45 μm Millipore filter and after this handling
about half of the volume of the samples were acidified with
concentrated HNO_3 to pH 2-3.

2. Analytical methods

2.1 *Soil samples*

Except if noted otherwise, details of the analytical
procedures are given by the "Methods of soil, rock and water
analyses" of the Laboratory of Physical Geography and Soil
Science (1971, 1974).

- *Granulometric analysis:* The fine earth fraction was
  penetrated with 9% (v/v) H_2O_2 and 0.1 M HCl in a
  boiling waterbath to remove cementing agents. Salts
  were removed by suction and washing. The salt and
  clay fraction was obtained by wet sieving of a

suspended fine earth. Peptization was carried out by Na-pyrophosphate 3 mmol/litre. The fractions < 2 μm and 2-50 μm were obtained by the pipetting method, using Stokes law. The fractions >50 μm were obtained by dry sieving.

- *Organic matter:* organic matter was oxidized in the air-dry state by $K_2Cr_2O_7$ in concentrated H_2SO_4 under steady heating up to $175^\circ C$ for 90 seconds. The colour intensity of the formed green chromo-ions is measured colorimetrically.

- *pH:* pH has been determined potentiometrically in a 25 ml aqua dest. c.q. 0.01 M $CaCl_2$-extract of 10 g fine earth using a glass-electrode and an calomel-electrode.

- *Separation of the clay fraction for elemental analysis:* the clay fraction of the fine earth fraction was separated after the oxidation of organic matter (not on waterbath) and dispersion with 4 M NaOH (to pH = 8). The clay separate was saturated with Li by equilibration with a 2 M LiCl (pH = 7) and dialyzed against distilled water until completely free of Cl ions, and the clay was recovered by freeze drying.

- *Dithionite-extractable iron, aluminium, manganese and silica:* 1 g of fine earth was mixed with 130 ml of 0.54 M Na-citrate, 2 g $Na_2S_2O_4$ and was shaken over-night (Method of Holmgren). Iron, aluminium, manganese and silica were estimated with an argon-plasma spectrometer.

- *Oxalate-extractable iron, aluminium, manganese and silica:* 2 g of fine earth was shaken in 100 ml of Tamm's solution ($COOH_2.2H_2O$ 21.0 g and NH_4-oxalate 28.4 g/litre) in the dark for 4 hours. Iron, aluminium, manganese and silica were estimated by an argon-plasma spectrometer.

- *Pyrophosphate-extractable iron, aluminium and manganese:* 2 g of fine earth was shaken overnight in 150 ml 0.1 M Na-pyrophosphate. Iron, aluminium and manganese were estimated by an argon-plasma spectrometer.

- *Exchangeable cations: Ca, Mg, Na and K:* 5.0 g of fine earth is shaken four times in a 25 ml 0.5 N LiCl/LiAc exchange solution. Ca and Mg were estimated colorimetrically with glycoxal bis (2-hydroxyanil) and with tital yellow. Na and K were estimated by flame photometry. *Al and H:* 1-10 g of fine earth is shaken 5 times in a 30 ml 1 N KCl solution. Al + H were estimated by titrating with 0.01 N NaOH solution after addition of a few drops of phenolphtalein until the colour of the solution is "permanently" pink during 1 minute. Al was estimated colorimetrically with pyrocatacholviolet.

- *CO_2 content:* CO_2 content was estimated after destruction of the carbonate by 4 N HCl in a weighted apparatus which is open to the air. Determination of the loss of weight after the reaction has ceased.

- *N content:* 250-1000 mg fine earth is put into a Kjeldahl flask. The organic matter is destructed with H SO and a Se-mixture. Steam-destillation of the ammonia, after liberation with excess alkali (NaOH) into a solution of boric acid and back-titration of the ammonium borate to boric acid with KH $(IO_3)_2$.

- *Elemental analysis of the rocks, fine earth and clay fraction:* total Al, Fe, Mn, Na, K, Ca, Mg, Ti and P were estimated after decomposing the material with HF and H_2SO_4. Total Al was estimated colorimetrically with pyrocatacholviolet; total Fe colorimetrically with orthophenanthroline; total Mn colorimetrically with formaldoxim; total Ca colorimetrically with glyoxal bis (2-hydroxyanil); total Mg colorimetrically with titan yellow; total Ti colorimetrically with tiron; total P colorimetrically as phosphorus-molydenum complex; total Na and K were estimated by flame photometry; total Li in the Li-saturated clay fraction was measured by flame photometry. For the Fe(II)-Fe(III) analysis, the rock, fine earth or clay fraction was decomposed at 60°C for 1 minute in a HF-H_2SO_4 mixture in a covered polyethylene bottle. After transferring the reaction mixture into a saturated boric acid solution, Fe(II) was measured colorimetrically with orthophenanthroline, and total iron was measured afterwards in the same solution upon adding a solid reductant (hydroquinone). Total Si was estimated after soda fusion as the blue coloured β-silico-mobybdenic acid.

- *X-ray diffraction analysis:* x-ray diffraction analyses on disoriented clay samples were carried out by means of a quadruple Guinier-de Wolff camera using Co Kα radiation. Diffractograms were obtained with a Philips vertical diffractometer with Cu Kα radiation using well oriented clay samples. Diffractograms were made of clay samples saturated with Mg, Mg-ethylene glycol, K, heated to 300°C and heated to 550°C.

2.2 Water samples

In the laboratory the field filtered samples were stored in the dark at 4°C.

- *pH:* pH was potentiometrically measured immediately after opening the polyethylene bottle with a combined glass-calomel electrode.

- *Alkalinity:* the alkalinity was estimated by potentiometric titration with 0.01 N H_2SO_4 to pH = 4.6.

181

- Cl: titrated with 0.02 N $Hg(NO_3)_2$ using diphenylcarbazone and bromphenol blue in ethanol as mixed indicator.

- $NO_3^- + NO_2^-$: nitrate plus nitrite was estimated as the NO_2-sulphanilic acid complex. Nitrate was reduced to nitrite by copperized cadmium in the presence of chloride. In the spring- and river waters only nitrate could be detected.

- SO_4^{2-}: sulphate was determined turbidemitrically as $BaSO_4$ stabilized by "Tween 80".

- $H_4SiO_4^0$: aqueous silica was estimated colorimetrically as the blue coloured β-silico-molybdenic acid, after adding ammonium molybate and a reductant $(Na_2SO_3 + metol + K_2S_2O_5)$ in a medium of sulphuric and tartaric acid.

- Na^+ and K^+: were determined by a Lange flame photometer (air-butane flame) using $Al(NO_3)_3$ to prevent interference of calcium

- Ca^{2+}: calcium was estimated colorimetrically with glyoxal bis (2-hydroxyanil)

- Mg^{2+}: magnesium was estimated colorimetrically with tital yellow

- Mn^{2+}: manganese was measured colorimetrically with formaldoxim

- Fe^{2+}: ferrous iron was determined colorimetrically with orthophenanthroline

- Al^{3+}: aluminium was estimated colorimetrically with pyrocatacholviolet

- NH_4^+: ammonium was determined colorimetrically by means of the indophenol-blue method

3. Selective iron, aluminium and manganese extractions

Much attention has been given to the nature and amounts of secondary iron-, aluminium- and manganese forms in soils (Schwertmann & Taylor 1977; Pa Ho Hsu 1977; Wada 1977; McKenzie 1977). These have been generally aimed at relating these compounds to soil forming processes or the classification of soils in temperate humid areas (McKeague & Day 1966; Blume & Schwertmann 1969; McKeague et al 1971; Duchaufour 1977). In this section a short review is given of the various extraction techniques for the secondary iron-, and aluminium- and manganese forms in soils (for methods see appendix I).

Iron

The various forms of iron present in soils can be listed as follows(see also Andriesse 1975):

a) Silicate bound iron and compound metal oxides such as ilmenite
b) Crystalline free iron: geothite and hematite

c) Crystalline free iron: magnetite, maghemite and
   lepidocrocite
d) Aged amorphous hydrous iron, including ferrihydrite
e) Amorphous hydrous iron - non-aged gels
f) Fulvic acid complexed iron
g) Humic acid complexed iron

The various iron compounds are more or less selectively
extracted from the soils as follows (see also table 34):

1. Pyrophosphate extractable iron (Fe_p) at pH 10:   the
   pyrophosphate extracts the fulvic and the humic acid
   complexed iron, and amorphous hydrous iron non-aged
   gels ((e) + (b) + (g)).

2. Oxalate extractable iron (Fe_o) at pH 3:   oxalate
   extracts the compounds mentioned under (d), (e), (f)
   and partly (c). If sources of ferrous iron such as
   siderite are present, considerable parts of (b) will be
   attacked.

3. Dithionite extractable iron (Fe_d): dithionite extracts
   the compounds mentioned under (b), (c) with only
   small amounts of magnetite, (d),(e),(f),(g).

4. Total extractable iron (Fe_t):   this value represents
   compounds mentioned under (a) to (g).

Since in the parent rocks in the Haarts catchment no
magnetite occurs, while maghemite and siderite are absent
and ilmenite and pyrite are only present as traces in the
soils, the following conclusions can be drawn:

   a. Fe_{t-d} values represent the silicate bound iron in
       the soils
   b. Fe_{d-o} values represent the well crystallised free
       iron oxides (geothite and hematite;   some-
       times lepidocrocite)
   c. Fe_{o-p} values represent the aged amorphous inorganic
       iron compounds
   d. Fe_p values represent the organic complexed iron and
       the non-aged amorphous iron gels ("mobile
       fraction")
   e. Fe_o/Fe_d ratio is a relative measure of the degree
       of ageing or crystallization of free iron
       oxides ("activity ratio")

Aluminium

For the aluminium compounds in soils it has to be remembered
that the various extractive techniques are less specific
than for iron. Nevertheless the dithionite extractable
(Al_d), oxalate extractable (Al_o) and pyrophosphate
extractable (Al_p) aluminium are used for characterising
soil horizons and soil forming processes. McKeague et
al (1971) showed that pyrophosphate extracts are somewhat
less specific for organic complexed aluminium and aluminium
hydroxide soils, because it also extracts somewhat amorphous
hydrated aluminium (20%). Oxalate extractable aluminium
(Al_o) appears to be more representative for amorphous
forms of aluminium than dithionite extractable aluminium

(Al$_d$) (McKeague & Day 1966). For tropical soils containing gibbsite Al$_d$ seems to be more representative, although gibbsite is only partly dissolved (Scheffer & Schachtschabel 1976; see also table 34). For soils in the temperate humid climate a fairly good correlation is obtained between the Al$_d$ and Al$_o$ values and the shape of the depth functions for these two Al fractions are almost identical (Schlichting & Blume 1969). This result is confirmed by the investigation of the soils in the Haarts catchment.

However, the Al$_d$ values for the soils in the investigated area were always somewhat higher than the Al$_o$ values (fig. 12, 13, 14 and 15). Gamble & Daniels (1972) found that dithionite extracted much more silica from kaolinite and fine-grained quartz than oxalate. Oxalate extracts contained less than 10 percent of the dithionite extractable amounts. Silica (as SiO_2) in the dithionite was also determined and ranged from 0.03 to 0.13 percent for the various soil horizons. If it is assumed that all the silica comes from a (clay) mineral with a SiO_2/Al_2O_3 ratio of 2 (kaolinite) the equivalent amounts of aluminium can be calculated. These amounts of aluminium with some exceptions are in the same order as the difference between the Al$_d$ and Al$_o$ levels might be tentatively explained by a slight attack of (poor crystalline) layer silicates by the dithionite extract and Al$_o$ appears to be perhaps a more realistic measure for the pedogenetic aluminium oxides than Al$_d$.

Manganese

For the various extraction techniques for manganese similar comments can be made as for aluminium. The "free manganese oxide" can be well separated from the silicate bound manganese by using the dithionite (Mn$_d$)- or oxalate (Mn$_o$) extractions. The silicate bound manganese (Mn$_{t-d}$) is obtained by subtracting the Mn$_d$ (or Mn$_o$) values from the total manganese content (Blume & Schwertmann 1969). If carbonates are present, which is not the case for the soils in the investigated area, the Mn$_{t-d}$ value (not Mn$_{t-o}$), because of the low pH in the oxalate extract) represents manganese in silicates and carbonates. Pyrophosphate extractable manganese (Mn$_p$) appears to indicate the organic complexed manganese, although this extraction is less specific than for iron.

Table 34 Dissolution of Al, Fe and Si in various soil components by treatment with different reagents (modified from Wada 1977)

Elements in specified component and complex	Treatment with			
	$0.1 \text{ M } Na_4P_2O_7$ (pH = 10)[x]	Diothionite-citrate[xx]	Acid NH_4-oxalate (pH = 3.0)[xxx]	0.5 M NaOH [xxxx]
Al in:				
Organic complexes	good	good	good	good
Hydrous oxides				
Non-crystalline	poor	good	good	good
Crystalline	no	poor	no	good
Fe in:				
Organic complexes	good	good	good	no
Hydrous oxides				
Non-crystalline	poor	good	good	no
Crystalline	no	good	no	no
Si in:				
Opaline silica	no	no	no	good
Crystalline silica	no	no	no	poor
Al and Si in:				
Allophanelike	poor	good	good	good
Allophane	poor	poor	good	good
Imogolite	poor	poor	good-fair	good
Layer silicates	no	no	no	poor-fair

x   Bascomb (1968): McKeague et al (1971); Wada & Higashi (1976)

xx   Mehra & Jackson (1960); Holmgren (1967)

xxx   Schwertmann (1964); McKeague et al (1971); Higashi & Ikada (1973); Wada & Wada (1976)

xxxx   Hashimoto & Jackson (1960); Wada & Greenland (1970); Tokashiki & Wada (1975)

APPENDIX 2

PROFILE DESCRIPTIONS AND
ANALYTICAL DATA ON THE SOIL PROFILES

PROFILE HAARTS 2

Classification: Lithic Udorthent (dystric regosol);
                 loamy-skeletal; mixed, mesic
Location: See fig 10
Elevation: 465 m
Land form: S facing upper part of valley spur
Slope: 1°
Vegetation: mixed oak-beech forest
Profile: moist throughout; roots till R1 horizon

Profile description (abbreviated)

O   2-0 cm   very dark brown (10 YR 2/2)
                       partly decomposed organic matter; abrupt
                       and smooth boundary

A11   0-17 cm   dark brown (7.5-10 YR 3/2)
                       gravelly loam; many angular shale and
                       quartzitic shale fragments; moderate
                       medium crumb; loose; many roots;
                       clear and smooth boundary

A12   17-22 cm   dark brown to dark yellowish brown
                       (7.5-10 YR 4/4); very gravelly and stony
                       loam; many angular shale and quartzitic
                       shale fragments; weak fine angular blocky;
                       friable; many roots; abrupt and smooth on

R1   22-35 cm   grayish brown (10 YR 5/2) slightly
                       disintegrated shale

R2   +35 cm   grayish brown (2.5 YR 5/2) very
                       slightly disintegrated shale

Table 35 Particle size distribution, organic carbon, nitrogen, C/N ratio

Depth (cm)	Hor	Particle size (mm)**		(mm)*				(μm)*		org.C*	N*	C/N ratio
		>2	2-1	1-0.5	0.5-0.25	0.25-0.1	0.1-0.05	50-2	<2			
1-11	A11	48.5	15.5	5.0	3.0	5.0	7.5	41.0	23.0	5.6	0.36	15.6
18-22	A12	54.0	9.5	3.0	3.0	6.0	8.0	45.5	25.0	2.35	0.16	14.8

* weight perc. abs. dry fine earth; ** weight perc. air dry soil

Table 36 pH and exchangeable cation characteristics

Depth (cm)	Hor	pH		Exchangeable cations in meq/100 g fine earth							Base sat.%
		H_2O	$CaCl_2$	Na	K	Ca	Mg	Al	H	= CEC	
1-11	A11	4.5	4.1	0.16	0.45	2.80	0.61	2.32	2.56	8.9	45.2
18-22	A12	4.1	3.6	0.14	0.19	0.37	0.06	5.26	3.60	9.6	7.8

Table 37 Extractable iron, aluminium and manganese by dithionite, ammonium oxalate and sodium pyrophosphate, in the fine earth fraction (weight perc.)*

Hor	A11	A12	Hor	A11	A12
Fe_d	3.48	3.25	Al_d	0.23	0.20
Fe_o	0.64	0.62	Al_o	0.23	0.20
Fe_p	0.17	0.14	Al_p	0.18	0.16
Fe_{t-d}	0.21	0.25	Mn_d	0.22	0.10
Fe_o/Fe_d	0.18	0.19	Mn_o	0.21	0.09
Fe_p/Fe_o	0.26	0.22	Mn_p	0.15	0.05
			Mn_{t-d}	0.02	0.01

Table 38   Elemental composition of the fine earth and clay
            fractions (weight perc.) and derived molar
            ratios

| | Fine earth | | Clay | | Rock* |
Hor	A11	A12	A11	A12	
SiO_2	63.9	67.8	49.5	49.8	67.3
Al_2O_3	14.2	15.5	28.6	28.9	17.8
Fe_2O_3	4.36	4.49	6.59	6.45	4.61
FeO	0.42	0.29	0.14	0.15	0.11
MnO	0.28	0.14	0.32	0.19	0.16
MgO	0.38	0.38	0.68	0.68	0.45
CaO	0.32	0.06	0.03	0.07	0.05
Na_2O	0.77	0.84	0.98	1.04	1.12
K_2O	2.34	2.52	4.70	4.67	3.34
TiO_2	0.96	1.08	0.88	0.88	0.91
P_2O_5	0.19	0.14	0.70	0.40	0.10
SO_3	0.21	0.20	-	-	0.20
Li_2O	-	-	0.23	0.30	-
LOI	11.50	6.52	6.66	6.44	3.81
SiO_2/R_2O_3	6.24	6.19	2.55	2.55	5.48
SiO_2/Al_2O_3	7.66	7.41	2.93	2.92	6.41
SiO_2/Fe_2O_3	33.65	37.46	19.51	20.01	37.81
Al_2O_3/Fe_2O_3	4.39	5.05	6.65	6.85	5.90

*very slightly altered rock

Table 39 Normative mineralogical composition of the fine earth, clay fractions and rock (weight perc.)

Hor	Fine earth A11	Fine earth A12	Clay A11	Clay A12	Rock**
Q*	48	48 (48)	8.2 (9.6)	11.3 (12.7)	43
Ab	7.1 (7.1)	7.4 (7.4)	8.3 (8.3)	9.0 (9.1)	9.5
K-mica/Ill	25 (25)	28 (28)	55.4 (55.5)	56.9 (56.9)	28
Kaol	8.8 (8.4)	9.3 (8.8)	16.0 (14.3)	10.7 (9.1)	11.5
Chlor	0.8 (1.4)	0.8 (1.4)	- (2.4)	- (2.4)	1.6
Fe-Chlor	1.5 (1.5)	0.6 (0.7)	- (0.3)	- (0.4)	0.3
Verm	0.5 (-)	0.6 (-)	2.2 (-)	2.3 (-)	-
Fe-Verm	0.1 (-)	0.1 (-)	0.3 (-)	0.4 (-)	-
Hm + Go	4.7 (4.7)	4.7 (4.7)	6.4 (6.4)	6.9 (6.9)	4.5
Ru	1.1 (1.1)	1.2 (1.2)	0.9 (0.9)	1.0 (1.0)	0.9
MnO$_2$	0.4 (0.4)	0.2 (0.2)	0.4 (0.4)	0.2 (0.2)	0.1
Apa	0.1 (0.1)	0.1 (0.1)	- (-)	0.1 (0.1)	0.1
Str	0.4 (0.4)	0.3 (0.3)	1.9 (1.9)	1.1 (1.1)	0.1
Pr	0.1 (0.1)	0.1 (0.1)	-	-	0.4
Zo	0.9 (0.9)	0.1 (0.1)	-	-	tr

*Q = free silica; Ab = albite; Ill = illite; Kaol = kaolinite; (Fe)-Chlor = chlorite; (Fe)-Verm = vermiculite; Hm = hematite; Go = geothite; Ru = rutile; Apa = apatite; Str = strengite; Pr = pyrite; Zo = zoisite; **very slightly altered rock

190

Table 40 Peak area percentages of X-ray diffractograms of the clay fraction, treated with Mg-ethylene glycol, and cation exchange capacity of the clay fraction after H_2O_2 treatment

Depth cm	Hor	d values ($\overset{\circ}{A}$)						CEC meq/100 g
		7	10	12	14	16	17-18	
1-11	A11	n.d.	90.5	n.d	5.5	2.5	1.5	15.3
18-22	A12	n.d.	87	n.d.	7.5	3.5	2	20

n.d. = not determined

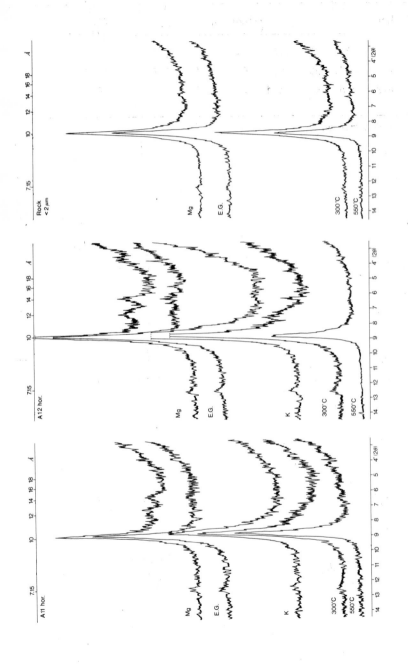

Figure 31. X-ray diffraction patterns for oriented clay separates of profile Haarts 2 (All, Al2, R horizon)

PROFILE NOTHUM 1

Classification:	Thapto-hapludultic Umbric Dystrochrept (dystric cambisol, overlying truncated acrisol); loamy-skeletål, mixed, mesic
Location:	50 m north of a point 700 m southwest of Nothum along country road to Bavigne (see fig. 10)
Elevation:	485 m
Slope:	5°
Vegetation:	mixed oak-beech coppice with undergrowth of blackberry
Profile:	moist throughout; roots throughout

Profile description (abbreviated)

O	2-0 cm	very dark brown (7.5 YR 2/2) partly decomposed organic matter- abrupt and smooth boundary
A1	0-10 cm	dark brown (7.5 YR 2.5/2) gravelly silt loam; many fresh angular shale and quartzitic shale fragments; moderate medium crumb; clear and smooth boundary
A3	10-26 cm	dark brown to brown (7.5 YR 3.5/2) gravelly silt loam; many fresh angular shale and quartzitic shale fragments; weak medium crumb; clear and irregular boundary
B2	26-40 cm	dark brown to brown (7.5 YR 4/4) gravelly silty clay loam with yellowish brown (10 YR 5/4) mottling; many fresh angular shale and quartzitic shale fragments; moderate very fine to moderate coarse subangular blocky; friable; clear and smooth boundary
IIB1tb	40-50 cm	yellowish brown (10 YR 5/6) gravelly and stony silty clay loam; moderate fine to medium subangular blocky; sticky; patchy thin ferriargillans; very many fresh shale and quartzitic shale fragments; abrupt and smooth boundary
IIB2tb	50-70 cm	light gray (2.5 Y 7/2) gravelly and stony silty clay loam, few strong brown (7.5 YR 5/6) mottles

NB Animal burrows up to 7 cm diameter filled with brown loam penetrate into B2 and IIBtb horizons

Micromorphological description - see table 41

Table 41 Micromorphological description – Nothum 1 (from Kwaad & Mücher 1977)

	Horizon: A 1	A 3	B	IIBt b
Lithorelicts	many shale and sandstone fragments (1–7 mm), fresh and with iron segregation, low and high sphericity, subangular to subrounded, no weathering rims	many rock fragments (2–10 mm), dominant shale, few sandstone, mostly with iron segregation, colours in incident light yellow, orange, bright red, no weathering rims, low and high sphericity, subrounded, sometimes rounded	many rock fragments (2–5 mm), dominantly shale, few sandstone, little weathered, low sphericity, angular	horizon consists chiefly of rock fragments (5–20 mm) shale and sandstone, weakly weathered, low sphericity, subrounded; weak parallel distribution, inclined to the surface
Humiskel	few to common coarse to extremely coarse root fragments with birefringence, few coarse red brown root fragments without birefringence	very few to few very coarse root fragments with birefringence	very few coarse and very coarse root fragments with and without very weak birefringence, very few medium root fragments with birefringence	very few very and extremely coarse root fragments with birefringence
Biorelicts	common charcoal fragments (up to 3.4 mm)	few charcoal fragments (coarse and very coarse sand size)	very few charcoal fragments (size of fine, medium and coarse sand)	not observed

Table 41 (continued)

Voids	many very fine, fine and meso-macro vughs; common (medium meso) craze planes; few (very fine macro) channels	many joint craze planes (mesopores) parallel to surface (chiefly at 15–20 cm depth); many vughs (medium and coarse mesopores, very fine macropores, coarse macropores); very few channels (very coarse macropores)	many craze planes (mesopores), common vughs (fine macropores) very few channels (coarse, meso and very fine macropores), locally compound packing voids (very fine macropores)	many skew planes (micro, meso and very fine macropores), many vughs (coarse meso and medium macropores), sometimes interconnected, few channels (medium and coarse mesopores)
S-matrix	porphyroskelic related distribution, skeleton grains (silt size) closely spaced	porphyroskelic related distribution	porphyroskelic related distribution	agglomeroplasmic related distribution with very porous plasma (intrapedal voids), locally intertextic
Skeleton grain	abundant silt and common very fine sand grains, dominantly quartz, high sphericity, subangular	abundant silt and common very fine sand grains, dominantly quartz, high and low sphericity, subangular	abundant silt and few very fine sand grains, dominantly quartz, high sphericity, subangular	common silt and very fine sand grains, dominantly quartz, high sphericity
Plasma type	mullicol, with a silasepic plasmic fabric, dark brown in incident light, due to organic matter	dominantly mullicol, silasepic; spots of argillicol, stained by iron, in lower part of horizon	argillicol, stained brown by iron	argillicol, partly grayish, partly brownish

Table 41 (continued)

	Horizon:		
	A 1	A 3	B	IIBt b

	A 1	A 3	B	IIBt b
Plasma re-orientations	not observed	not observed, except in an aggregate at 23 cm depth (mo-masepic plasmic fabric)	fine weak insepic plasmic fabric, strong fine masepic plasmic fabric at borders of some fecal pellets	greyish plasma: strong, very coarse wavy masepic plasmic fabric (matrans?) brownish plasma. clear mosepic plasmic fabric, fine and medium around some litho-relicts: medium skel-sepic plasmic fabric.
Cutans	not observed	not observed	one single fine illu-viation (?) ferri-argillan at 39 cm depth	common medium, yellow plane and channel illuviation ferri-argillans, common very thick compound illuviation cutans, composed of yellow ferriargillans, alter-nating with grey matriargillans, very few thick greyish illuviation matri-argillans

196

Table 41 (continued)

Glaebules	very few coarse to very coarse ferric nodules with rather diffuse boundaries	very few to few coarse and very coarse brown ferric nodules with diffuse and sharp boundaries, with and without skeleton grains; very few fine strong continuous yellow argillaceous papules in an aggregate at 23 cm depth	few dark brown ferric nodules with sharp boundaries and enclosed skeleton grains (medium to very coarse); very few very coarse ferric concretions with diffuse boundaries	iron segregation frequently occurs; very few strong continuous coarse papules
Pedotubules	few very coarse (0.4 mm) ortho aggrotubules and very few extremely coarse (2.1 mm) ortho-aggrotubules	very few to few very and extremely coarse ortho-aggrotubules	one very coarse ortho-striotubule at 38 cm depth	one meta tubule (3 cm) partly aggro-, partly isotubule, material derived from overlying horizon
Fecal pellets	many to abundant, single, sometimes welded, medium matric fecal pellets; few to common, mostly welded, very and extremely coarse (up to 1.3 mm) matric fecal pellets	many brown and dark brown medium matric fecal pellets, single, sometimes welded, in clusters and in channels; few very coarse or extremely coarse strongly welded matric fecal pellets	locally very coarse matric fecal pellets, welded	not observed

Table 42 Particle size distribution, organic carbon, nitrogen, C/N ratio

Depth	Hor	Particle size								org.c[*]	N[*]	C/N ratio
(cm)		(mm)[**]	(mm)[*]					(μm)[*]				
		> 2	2-1	1-0.5	0.5-0.25	0.25-0.1	0.1-0.05	50-2	<2			
1-10	A1	29.7	0.9	0.6	1.0	3.5	4.5	64.0	25.5	4.65	0.37	12.6
15-25	A3	31.0	2.5	1.5	2.0	4.0	5.0	60.0	25.5	2.5	0.23	11.0
30-40	B2	41.0	1.5	0.7	1.0	3.5	5.0	60.5	28.5	0.9	0.11	8.3
40-50	IIB1tb	34.5	0.7	0.6	0.9	2.5	4.0	63.5	28.0	0.4	0.09	4.1
50-70	IIB2tb	48.6	0.6	0.5	0.6	1.5	3.0	63.5	30.0	0.03	0.08	4.1

[*]weight perc. abs. dry fine earth; [**]weight perc. air dry soil

Table 43 pH and exchangeable cations characteristics

Depth	Hor	pH		Exchangeable cations in meq/100 g fine earth								Base
(cm)		H_2O	$CaCl_2$	Na	K	Ca	Mg	Al	H	CEC[*]	CEC[**]	sat. %
1-10	A1	4.1	3.9	tr	0.02	0.39	0.23	5.62	0.87	14.4		9.0
15-25	A3	4.3	3.9	tr	0.02	0.13	0.11	4.44	0.51	10.5		5.1
30-40	B2	4.3	3.9	0.01	0.01	0.08	0.08	3.36	0.37	7.8	7.1	4.6
40-50	IIB1tb	4.4	3.8	-	0.01	0.11	0.08	3.09	0.46	6.4	6.2	5.3
50-70	IIB2tb	4.4	3.8	-	0.01	0.25	0.17	4.25	0.41	6.9	7.2	8.4

[*]CEC - LiCl/Li acetate (0.5 N, pH=7.0); [**]CEC - NH_4 acetate (1.0 N, pH=7.0).

Table 44 Extractable iron, aluminium and manganese by dithionite, ammonium oxalate and sodium pyrophosphate in the fine earth fraction (weight percentages)[*]

Depth (cm)	Hor	Fe_d	Fe_o	Fe_p	Fe_{t-d}	Fe_o/Fe_d	Fe_p/Fe_o	Al_d	Al_o	Al_p
0-10	A1	3.17	0.96	0.30	0.22	0.30	0.31	0.43	0.38	0.34
15-25	A3	3.13	0.97	0.24	0.32	0.31	0.25	0.41	0.35	0.27
30-40	B2	2.57	0.62	0.18	0.52	0.24	0.29	0.33	0.26	0.17
40-50	IIb1tb	2.57	0.62	0.17	0.58	0.24	0.27	0.29	0.25	0.16
50-70	IIB2tb	3.02	0.57	0.11	1.22	0.19	0.19	0.26	0.21	0.10

(continued)

Depth (cm)	Hor	Mn_d	Mn_o	Mn_p	Mn_{t-d}
0.10	A1	0.17	0.16	0.10	0.04
15-25	A3	0.15	0.14	0.05	0.03
30-40	B2	0.12	0.10	0.02	0.02
40-50	IIb1tb	0.10	0.09	0.02	0.01
50-70	IIB2tb	0.03	0.03	tr	0.02

x organic matter free and abs. dry base

Table 45 Elemental composition of the Li-saturated clay fraction (weight perc.[*]) and derived molar ratios

Depth (cm)	Hor	SiO_2	Al_2O_3	Fe_2O_3	FeO	MnO	MgO	CaO	Na_2O	K_2O
0-10	A1	47.3	28.8	7.48	0.41	0.11	0.85	0.04	0.75	4.18
15-25	A3	46.6	29.2	7.90	0.36	0.12	0.90	0.07	0.78	4.25
30-40	B2	45.8	29.9	8.21	0.31	0.11	0.96	0.02	0.79	4.30
40-50	IIB1tb	45.7	29.7	8.19	0.40	0.12	1.02	0.01	0.81	4.34
50-70	IIB2tb	44.8	30.1	8.83	0.62	0.07	1.16	0.01	0.75	4.79

(continued)

Depth (cm)	Hor	TiO_2	P_2O_5	Li_2O	LOI	SiO_2/R_2O_3	SiO_2/Al_2O_3	SiO_2/Fe_2O_3	Al_2O_3/Fe_2O_3
0-10	A1	0.92	0.30	0.26	8.60	2.37	2.79	15.85	5.68
15-25	A3	0.93	0.29	0.34	8.24	2.29	2.71	14.93	5.52
30-40	B2	0.92	0.37	0.20	8.14	2.20	2.60	14.23	5.48
40-50	IIB1tb	0.90	0.32	0.35	8.17	2.20	2.61	14.07	5.39
50-70	IIB2tb	0.82	0.24	0.16	7.58	2.10	2.52	12.51	4.96

[*]abs. dry base

Table 46 Normative mineralogical composition of the clay fraction (weight perc.)

Depth (cm)	Hor	Q	Ab	Ill	Kaol	Chlor	Fe-Chlor
0-10	A1	6.9	6.4	50	23	-	-
15-25	A3	5.3	6.6	51	24	-	-
30-40	B2	3.7	6.6	51	24	-	-
40-50	IIB1tb	3.6	7.0	57	23	-	-
50-70	IIB2tb	{ 1.9	6.4	57	19	-	-
		(4.5)	(6.4)	(57)	(16)	(4.0)	(1.5)

Table 46 *(continued)*

Depth (cm)	Hor	Verm	Fe-Verm	Go	Ru	MnO$_2$	Str	Apa
0-10	A1	2.8	0.9	8.1	0.9	0.1	0.8	0.1
15-25	A3	2.9	0.8	8.5	0.9	0.2	0.6	0.1
30-40	B2	3.1	0.7	8.8	0.9	0.1	0.9	-
40-50	IIB1tb	3.3	0.9	8.8	0.9	0.2	0.8	-
50-70	IIB2tb	3.7	1.4	9.5	0.8	0.1	0.6	-
		(-)	(-)	(9.5)	(0.8)	(0.1)	(0.6)	(-) }

Table 47 Elemental composition of the fine earth fraction (weight perc.[x]) and derived molar ratios

Depth (cm)	Hor	SiO$_2$	Al$_2$O$_3$	Fe$_2$O$_3$	FeO	MnO	MgO	CaO	Na$_2$O	K$_2$O
0-10	A1	66.5	12.7	3.84	0.53	0.24	0.38	0.17	0.58	2.13
15-25	A3	68.9	13.4	4.28	0.39	0.22	0.40	0.19	0.62	2.21
30-40	B2	71.1	13.9	4.34	0.26	0.17	0.46	0.11	0.66	2.37
40-50	IIB1tb	70.0	15.1	4.42	0.30	0.15	0.53	0.13	0.67	2.56
50-70	IIb2tb	65.1	17.3	5.55	0.61	0.07	0.77	0.11	0.53	3.24

(continued)

Depth (cm)	Hor	TiO$_2$	P$_2$O$_5$	SO$_3$	LOI	SiO$_2$/R$_2$O$_3$	SiO$_2$/Al$_2$O$_3$	SiO$_2$/Fe$_2$O$_3$	Al$_2$O$_3$/Fe$_2$O$_3$
0-10	A1	0.91	0.14	0.15	11.81	7.29	8.92	39.90	4.47
15-25	A3	0.96	0.12	0.09	8.25	7.15	8.75	38.87	4.44
30-40	B2	0.97	0.11	0.11	5.51	7.17	8.69	40.85	4.70
40-50	IIB1tb	0.99	0.11	0.10	5.02	6.56	7.88	39.12	4.96
50-70	IIB2tb	0.99	0.10	0.10	5.49	5.19	6.38	27.80	4.36

[x]abs. dry base

Table 48 Normative mineralogical composition of the fine earth fraction (weight perc.)

Depth (cm)	Hor	Q	Ab	K-mica	Kaol	Chlor	Fe-Chlor	Verm
0-10	A1	54	5.4	24	7.6	0.7	1.2	0.7
15-25	A3	53	5.6	23	8.9	0.7	0.9	0.7
30-40	B2	53	5.6	25	8.3	0.7	0.4	0.9
40-50	IIB1tb	50	5.7	26	9.4	0.9	0.2	0.9
50-70	IIB2tb	43	4.6	33	8.3	1.5	1.0	1.1
		(44)	(4.6)	(33)	(7.3)	(2.7)	(1.5)	(-)

(continued)

Depth (cm)	Hor	Fe-Verm	Hm+Go	Ru	MnO_2	Str	Apa	Pr	Zo
0-10	A1	0.2	4.3	1.0	0.3	0.2	0.2	0.2	-
15-25	A3	0.2	4.6	1.0	0.3	0.2	0.2	0.1	-
30-40	B2	0.2	4.6	1.0	0.2	0.2	0.1	0.1	0.4
40-50	IIB1tb	0.25	4.7	1.0	0.2	0.2	0.1	0.1	0.4
50-70	IIB2tb	0.4	5.9	1.0	0.1	0.2	0.1	0.1	0.3
		(-)	(5.9)	(1.0)	(0.1)	(0.2)	(0.1)	(0.1)	(0.3)

Table 49 Peak area percentage of the X-ray diffractograms of the clay fraction, treated with Mg-ethylene glycol, and cation exchange capacity of the clay fractions after H_2O_2 treatment

Depth (cm)	Hor	d values (Å)						CEC meq/100 g
		7	10	12	14	16	17-18	
0-10	A1	n.d.	78	n.d.	9	9	4	17.3
15-25	A3	n.d.	74	n.d.	11	9	6	22.7
30-40	B2	n.d.	81	n.d.	11	5	3	20.3
40-50	IIB1tb	n.d.	81	n.d.	10	6	3	23.3
50-70	IIB2tb	n.d.	89	n.d.	6	3	2	10.7

n.d. = not determined

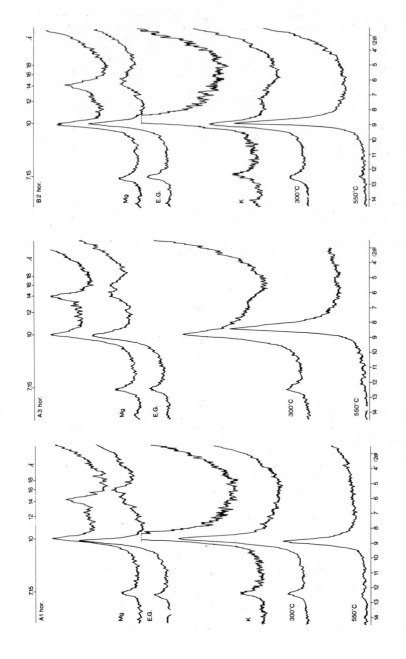

Figure 32a X-ray diffraction patterns for oriented clay separates of profile Nothum 1
(A1, A3, B2 horizon)

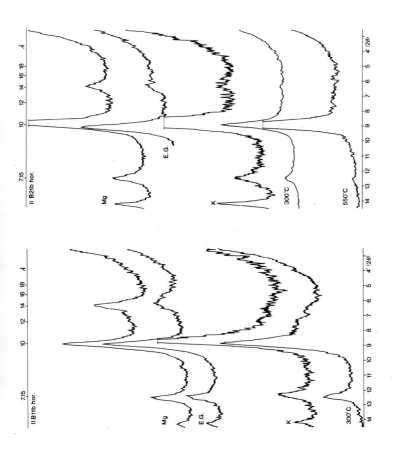

<u>Figure 32b</u> X-ray diffraction patterns for oriented clay separates of profile Nothum 1
(IIBltb and IIB2tb horizon)

203

PROFILE NOTHUM 3 (from Kwaad & Mücher 1977)

Classification: Umbric Dystrochrept (dystric cambisol);
loamy-skeletal, mixed, mesic

Location: 50 m south of a point at 1.4 km southwest
of Maison Schuman along road N 26 to
Bavigne; ± 6 km southwest of Wiltz, north
Luxembourg (see fig. 10)

Elevation: 433 m

Land form: north facing valley slope (lower slope)
of a tributary of the Birbaach river

Slope: 17°

Vegetation: mixed oak-beech forest, poorly developed

Profile: moist throughout, roots throughout

Profile description (abbreviated)

A1   0-10 cm   dark brown (7.5 YR 3/2) gravelly loam,
many fresh angular shale and sandstone
fragments; moderate medium crumb; clear
wavy boundary

A3   10-31 cm   brown (7.5 YR 4/4) gravelly loam, many
fresh angular shale and sandstone
fragments; moderate medium and coarse
crumb; clear wavy boundary

B   31-45 cm   yellowish brown (10 YR 5/6) very gravelly
and stony loam, very many more or less
flat shale and sandstone fragments (up to
25 cm), mostly parallel to the surface;
moderate medium subangular blocky;
abrupt wavy boundary

IIC + R1 + 45 cm   olive yellow (2.5 Y 6/6) loamy gravel and
stones, dominant shale and sandstone
fragments

NB   Roman II numeral is based upon certain micromorphological
characteristics charcoal fragments, para-aggregates of
the uppermost 45 cm of the profile

Micromorphological description - see table 50

Table 50 Micromorphological description – Nothum 3 (Kwaad & Mücher 1977)

	Horizon: A 1	A 3	B
Lithorelicts	many shale and sandstone fragments (2–10 mm), colours in incident light gray, yellow, orange, red, high and low sphericity, sub-angular, subrounded and rounded	same as A 1 horizon	same as A 1 horizon
Humiskel	many very coarse root fragments with birefringence	common medium and coarse root fragments with birefringence	few to common medium and coarse, sometimes very coarse root fragments with bire-fringence (sometimes very weak)
Biorelicts	common charcoal fragments (up to 4.2 mm)	same as A 1 horizon	few charcoal fragments (size of medium and very coarse sand)
Voids	abundant compound packing voids (macro pores), common joint craze planes parallel to the surface from 3 to 6.5 cm depth (meso pores), few channels (very fine macro pores)	abundant compound packing voids (macro pores), few planes (fine and medium meso pores), few channels (fine meso and very fine macro pores), common vughs in aggregates (meso pores)	same as A3 horizon, maximum porosity from 31 to 36 cm depth
S-matrix	the fine earth of the horizon mainly consists of fecal pellets	the fine earth of the horizon mainly consists of loose aggregates with a porphyroskelic related distribution pattern	same as A 3 horizon

206

Table 50 (continued)

Skeleton grains	many silt and few to common very fine sand grains, dominantly quartz, fewer mica's, high sphericity (quartz), low sphericity (mica's), subangular to sub-rounded, occur in aggregates and between fecal pellets	many fine silt (mica's) and silt grains, common very fine sand and few fine sand grains, mainly quartz, high (quartz) and low (mica) sphericity, sub-angular, occur in aggregates	same as A 3 horizon
Plasma type	no plasma observed	mullicol in aggregates with brown staining by organic matter	argillicol in aggregates, with brown iron staining
Plasma reorientations	not observed	not observed (silasepic plasmic fabric)	not observed (silasepic plasmic fabric)
Cutans	not observed	not observed	not observed
Glaebules	not observed	not observed	not observed
Pedotubules	not observed	few extremely coarse ortho- and meta-aggrotubules	same as A 3 horizon
Fecal pellets	abundant medium dark brown single or weakly welded, sometimes strongly welded organic fecal pellets	many to abundant coarse and very coarse matric fecal pellets, single to strongly welded, very few medium single organic fecal pellets, mostly in clusters	not observed
Aggregates	not observed	many mullicol aggregates	many argillicol aggregates

207

PROFILE HAARTS 1

Classification:	Thapto-hapludalfic Umbric Dystrochrept (dystric cambisol, overlying a luvisol); loamy-skeletal, mixed, mesic	
Location:	see fig. 10	
Elevation:	405 m	
Land form:	ESE facing lower valley slope of the Haarts brook	
Slope:	22°	
Vegetation:	mixed oak-beech forest	
Profile:	moist throughout, roots till III?R1+c horizon	

Profile description (abbreviated)

O	1-0 cm	very dark brown (10 YR 2/2) partly decomposed organic matter; abrupt on smooth boundary
A11	0-4/5 cm	dark brown (10 YR 3/3) gravelly loam, many angular shale and quartzitic shale fragments; strong fine granular; friable; clear and smooth boundary
A12	4/5-7/8 cm	dark brown (7.5 YR 3.5/3) gravelly loam, many angular and quartzitic shale fragments; strong fine subangular blocky; friable; clear and wavy boundary
B2	7/8-31/33 cm	brown to strong brown (7.5 YR 5/5) gravelly and stony loam, many fresh shale fragments; moderate coarse subangular blocky disintegrating into weak fine subangular blocky; friable; clear and wavy boundary
B3	.31/33-55/60 cm	brown (7.5 YR 5/4) gravelly and stony silt loam, many angular shale fragments; weak coarse subangular blocky, disintegrating into weak fine subangular blocky; friable; clear and wavy boundary
B3+R1	55/60-85/90 cm	yellowish brown (10 YR 5/8) very gravelly and stony silt loam, very many angular shale fragments; weak subangular blocky; plastic; gradual and wavy boundary
IIB2tb	85/90-170 cm	yellowish brown (10 YR 5/6) very gravelly and stony silt loam; many to very many angular shale fragments; weak fine subangular blocky; friable to firm; patchy thin cutans; gradual and wavy boundary (samples 110-120 cm, 140-150 cm, 160-170 cm)

IIC + R1 170-200 cm   yellowish brown (10 YR 5/6) very gravelly and stony loam, very many angular shale fragments; massive; gradual and wavy boundary

III?R1+C 200-220 cm   yellowish brown (10 YR 5/6) silty loamy gravel and stones (angular shale fragments); clear and wavy boundary

IV?R1 200-240 cm   grayish brown (10 YR 5/2) slightly disintegrated shale fragments

IVR2 + 240 cm   olive gray (5 Y 5/2) hard (unaltered) shale, somewhat fractured

Table 51 Particle size distribution, organic carbon, nitrogen, C/N ratio

Depth (cm)	Hor	Particle size (mm)**		(mm)*				(µm)*		org.C*	N*	C/N ratio
		> 2	2-1	1-0.5	0.5-0.25	0.25-0.1	0.1-0.05	50-2	< 2			
0-7	A1	45.4	19.0	4.5	2.5	2.0	3.5	48.5	20.5	7.2	0.45	16.0
12-25	B2	49.8	16.5	4.5	3.0	2.5	3.5	49.0	22.0	2.0	0.14	14.1
40-50	B3	51.9	13.0	5.0	3.5	3.5	4.0	50.5	21.0	1.1	0.08	13.3
65-80	B3t+R1	74.5	10.5	5.0	5.0	4.0	4.0	55.5	16.0	0.55	0.07	7.9
110-120	IIB2tb	55.0	13.0	6.0	6.0	5.5	5.0	46.0	18.5	0.3	0.08	4.1
140-150	IIB2tb	56.2	12.0	5.5	6.5	6.5	6.5	46.0	19.0	0.3	0.09	2.9
160-170	IIB2tb	57.8	12.0	7.0	7.5	5.5	4.0	42.5	21.5	0.25	0.09	2.8
175-185	IIC+R1	73.5	15.0	8.0	9.0	6.5	4.0	35.5	22.0	0.35	0.09	4.1
200-220	III?R1+C	81.1	18.0	12.0	12.0	7.5	4.0	32.0	16.0	0.3	0.09	3.4
220-240	IV?R1	76.7	14.0	8.5	11.5	10.0	6.0	31.5	18.0	0.3	0.09	3.6

*weight perc. abs. dry fine earth; **weight perc. airdry soil

Table 52 pH and exchangeable cations characteristics

Depth (cm)	Hor	pH H_2O	$CaCl_2$	Na	K	Ca	Mg	Al	H	Σ=CEC	CEC*	Base sat.%
0-7	A1	3.95	3.8	0.22	0.37	0.40	0.27	5.13	3.69	10.1		12.5
12-25	B2	4.35	3.9	0.18	0.16	0.05	0.10	3.73	2.35	6.6		7.5
40-50	B3	4.55	3.95	0.18	0.16	0.06	0.30	2.99	2.09	5.8	6.0	12.1
65-80	B3+R1	4.7	4.0	0.16	0.15	0.07	0.80	2.30	1.92	5.4	5.1	21.9
110-120	IIB2tb	5.1	4.0	0.10	0.13	1.18	2.30	1.25	1.16	6.1	5.8	60.6
140-150	IIB2tb	5.25	3.95	0.09	0.10	1.34	2.22	0.97	0.86	5.7	6.2	66.0
160-170	IIB3tb	5.1	3.9	0.07	0.09	1.02	2.45	1.74	0.72	6.1	6.0	59.6
175-185	IIC+R1	5.05	3.9	0.07	0.09	0.79	2.62	1.83	0.48	5.9		60.7
200-220	III?R1+C	5.4	4.1	0.08	0.11	0.87	2.68	1.19	0.45	5.4		69.6
220-240	IV?R1	5.5	4.3	0.06	0.12	1.20	3.15	0.27	0.50	5.3		85.5

*CEC – NH_4 acetate (1.0 N, pH = 7.0)

Table 53   Extractable iron, aluminium and manganese by dithionite, ammonium oxalate and sodium pyrophosphate in the fine earth fraction (weight percentages[*])

Depth (cm)	Hor	Fe_d	Fe_o	Fe_p	Fe_{t-d}	Fe_o/Fe_d	Fe_p/Fe_o
0-7	A1	3.36	0.49	0.25	0.40	0.15	0.51
12-15	B2	3.31	0.41	0.10	0.43	0.12	0.43
40-50	B3	3.05	0.28	0.11	0.59	0.09	0.41
65-80	B3+R1	3.20	0.20	0.07	0.57	0.06	0.35
110-120	IIB2tb	3.70	0.30	0.03	0.12	0.08	0.10
140-150	IIB2tb	4.03	0.38	0.03	0.10	0.09	0.08
160-170	IIB2tb	4.16	0.51	0.02	0.11	0.12	0.04
175-185	IIC+R1	4.97	0.17	0.02	0.11	0.03	0.08
200-220	III?R1+C	4.80	0.17	0.01	0.12	0.04	0.08
220-240	IV?R1	4.62	0.15	0.01	0.13	0.03	0.07

(continued)

Depth (cm)	Hor	Al_d	Al_o	Al_p	Mn_d	Mn_o	Mn_p	Mn_{t-d}
0-7	A1	0.30	0.29	0.23	0.18	0.17	0.14	0.03
12-15	B2	0.25	0.25	0.14	0.10	0.09	0.03	0.02
40-50	B3	0.21	0.21	0.11	0.09	0.09	0.01	0.03
65-80	B3+R1	0.15	0.15	0.07	0.06	0.06	0.01	0.03
110-120	IIB2tb	0.12	0.08	0.04	0.09	0.09	0.01	0.02
140-150	IIB2tb	0.10	0.07	0.03	0.11	0.10	0.01	0.02
160-170	IIB2tb	0.10	0.07	0.03	0.10	0.10	0.01	0.01
175-185	IIC+R1	0.13	0.07	0.04	0.17	0.15	0.01	0.02
200-220	III?R1+C	0.14	0.09	0.03	0.23	0.22	0.01	0.02
220-240	IV?R1	0.13	0.08	0.03	0.25	0.24	0.01	0.01

[*] organic matter free and abs. dry base

Table 54 Elemental composition of the Li-saturated clay fraction (weight perc.[x]) and derived molar ratios

Depth (cm)	Hor	SiO_2	Al_2O_3	Fe_2O_3	FeO	MnO	MgO	CaO	Na_2O	K_2O
0-7	A1	49.7	27.0	7.57	0.39	0.37	0.82	0.03	0.82	3.82
12-25	B2	48.5	28.2	7.51	0.45	0.27	0.91	0.06	0.89	4.04
40-50	B3	49.3	27.7	7.43	0.46	0.25	0.95	0.07	0.85	4.01
65-80	B3+R1	49.0	27.9	7.50	0.48	0.24	1.03	0.04	0.90	4.43
110-120	IIB2tb	48.5	27.6	8.37	0.10	0.33	1.01	0.08	1.07	4.78
140-150	IIB2tb	48.7	28.4	7.75	0.08	0.30	0.73	0.05	1.28	5.37
160-170	IIB2tb	48.9	29.0	7.05	0.07	0.23	0.61	0.05	1.31	5.51
175-185	IIC+R1	48.1	29.1	7.34	0.05	0.42	0.67	0.04	1.20	5.41
200-220	III?R1+C	48.2	28.7	7.52	0.04	0.52	0.73	0.06	1.19	5.27
220-240	IV?R1	47.9	29.1	7.82	0.06	0.29	0.71	0.05	1.17	5.36

(continued)

Depth (cm)	Hor	TiO_2	P_2O_5	Li_2O	LOI	SiO_2/R_2O_3	SiO_2/Al_2O_3	SiO_2/Fe_2O_3	Al_2O_3/Fe_2O_3
0-7	A1	0.92	0.64	0.30	7.61	2.62	3.12	16.51	5.29
12-25	B2	0.91	0.39	0.32	7.57	2.47	2.92	16.10	5.51
40-50	B3	0.95	0.35	0.31	7.31	2.55	3.02	16.51	5.47
65-80	B3+R1	1.04	0.28	0.27	6.90	2.52	2.98	16.22	5.44
110-120	IIB2tb	0.80	0.25	0.24	6.87	2.49	2.98	15.20	5.10
140-150	IIB2tb	0.80	0.22	0.22	6.15	2.48	2.91	16.52	5.67
160-170	IIB2tb	0.68	0.22	0.21	6.13	2.47	2.86	18.42	6.38
175-185	IIC+R1	0.74	0.23	0.19	6.52	2.42	2.81	17.29	6.15
200-220	III?R1+C	0.76	0.25	0.10	6.70	2.44	2.85	16.94	5.94
220-240	IV?R1	0.72	0.25	0.09	6.52	2.38	2.79	16.15	5.78

[x]abs. dry base

Table 55 Normative mineralogical composition of the clay fraction (weight perc.)

Depth (cm)	Hor	Q	Ab	Ill	Kaol	Chlor	Fe-Chlor	Verm	Fe-Verm	Go	Ru	MnO_2	Str	Apa
0-7	A1	11.3	7.0	45	22	-	-	2.6	0.9	7.6	0.9	0.4	1.7	0.1
12-25	B2	8.3	7.5	48	22	-	-	2.9	1.0	7.9	0.9	0.3	0.9	0.1
40-50	B3	9.5	7.2	47	22	-	-	3.1	1.0	8.0	1.0	0.3	0.8	0.1
65-80	B3+R1	8.6	7.5	53	17	-	-	3.3	1.1	8.0	1.0	0.3	0.6	0.1
110-120	IIB2tb	7.7 (9.6)	9.0 (9.0)	57 (57)	12.0 (9.8)	- (3.5)	- (0.2)	3.3 (-)	0.2 (-)	9.2 (9.2)	0.8 (0.8)	0.4 (0.4)	0.5 (0.5)	0.1 (0.1)
140-150	IIB2tb	7.5	10.9	64	5.2	2.5	0.2	-	-	8.4	0.8	0.4	0.5	0.1
160-170	IIB2tb	6.7	11.1	65	5.5	2.1	0.2	-	-	7.5	0.7	0.3	0.5	0.1
175-185	IIC+R1	6.4	10.2	64	6.9	2.3	0.1	-	-	8.0	0.7	0.5	0.5	0.1
200-220	III?R1+C	7.1	10.1	63	7.6	2.5	0.1	-	-	8.1	0.8	0.6	0.5	0.1
220-240	IV?R1	6.3	9.9	63	7.7	2.5	0.1	-	-	8.4	0.7	0.4	0.5	0.1

Table 56 Elemental composition of the fine earth fraction (weight perc.[x]) and derived molar ratios

Depth (cm)	Hor	SiO_2	Al_2O_3	Fe_2O_3	FeO	MnO	MgO	CaO	Na_2O	K_2O
0-7	A1	63.1	13.8	3.83	0.78	0.23	0.45	0.11	0.76	2.22
12-25	B2	68.4	15.0	4.58	0.52	0.16	0.56	0.10	0.83	2.44
40-50	B3	69.4	15.3	4.59	0.46	0.16	0.61	0.11	0.85	2.50
65-80	B3+R1	68.8	15.8	4.87	0.43	0.11	0.63	0.10	0.88	2.99
110-120	IIB2tb	67.2	17.0	5.23	0.18	0.14	0.54	0.11	0.97	3.41
140-150	IIB2tb	65.5	17.9	5.80	0.07	0.17	0.41	0.04	0.99	3.70
160-170	IIB2tb	61.9	20.2	5.97	0.09	0.14	0.47	0.09	1.06	4.21
175-185	IIC+R1	60.8	19.9	7.15	0.07	0.24	0.48	0.04	1.04	4.28
200-220	III?R1+C	60.5	20.1	6.91	0.07	0.34	0.56	0.03	1.06	4.19
220-240	IV?R1	60.3	20.3	6.65	0.08	0.36	0.59	0.02	1.08	4.39

(continued)

Depth (cm)	Hor	TiO_2	P_2O_5	SO_3	LOI	SiO_2/R_2O_3	SiO_2/Al_2O_3	SiO_2/Fe_2O_3	Al_2O_3/Fe_2O_3
0-7	A1	0.88	0.18	0.18	13.60	6.39	7.78	35.72	4.59
12-25	B2	1.01	0.11	0.16	6.14	6.36	7.77	35.26	4.54
40-50	B3	1.01	0.11	0.18	4.76	6.35	7.71	36.17	4.69
65-80	B3+R1	0.96	0.09	0.16	4.14	6.07	7.37	34.20	4.64
110-120	IIB2tb	0.91	0.11	0.03	4.22	5.57	6.71	32.90	4.90
140-150	IIB2tb	0.94	0.12	0.03	4.38	5.14	6.22	29.63	4.76
160-170	IIB2tb	0.96	0.15	0.05	4.72	4.36	5.20	27.10	5.22
175-185	IIC+R1	0.95	0.12	0.03	4.97	4.21	5.19	22.35	4.30
200-220	III?R1+C	0.93	0.14	0.02	5.06	4.18	5.10	23.03	4.52
220-240	IV?R1	0.95	0.15	0.02	5.14	4.15	5.03	23.77	4.72

[x] abs. dry base

Table 57   Normative mineralogical composition of the fine
           earth fraction (weight perc.)

Depth (cm)	Hor	Q	Ab	K-mica	Kaol	Chlor	Fe-Chlor	Verm
0-7	A1	49	7.2	24	10.1	1.2	1.9	0.5
12-25	B2	48	7.2	24	10.1	1.2	1.1	0.6
40-50	B3	49	7.1	24	9.8	1.9	1.0	0.7
65-80	B2t+R1	47	7.6	28	7.4	1.7	0.9	0.5
110-120	IIB2tb	44 (44)	8.3 (8.3)	32 (32)	6.9 (6.5)	0.2 (1.9)	0.4 (0.4)	0.6 (-)
140-150	IIB2tb	41	8.7	35	6.5	1.4	0.3	-
160-170	IIB3tb	34	9.1	40	7.2	1.6	0.2	-
175-185	IIC+R1	34	9.0	41	5.7	1.7	0.2	-
200-220	III?R1+C	33	9.2	39	8.0	1.9	0.3	-
220-240	IV?R1	33	9.2	41	6.1	2.1	0.3	-

(continued)

Depth (cm)	Hor	Fe-Verm	Hm+Go	Ru	MnO_2	Str	Apa	Pr	Zo
0-7	A1	0.2	4.3	1.1	0.2	0.3	0.2	0.1	0.1
12-25	B2	0.2	4.9	1.1	0.1	0.2	0.1	0.1	0.2
40-50	B3	0.2	4.8	0.9	0.1	0.2	0.1	0.1	0.2
65-80	B3t+R1	0.2	5.0	0.9	0.1	0.1	0.1	0.1	0.1
110-120	IIB2tb	0.04 (-)	5.5 (5.5)	0.9 (0.9)	0.2 (0.2)	0.1 (0.1)	0.2 (0.2)	0.1 (0.1)	-
140-150	IIB2tb	-	5.9	0.9	0.2	0.1	0.1	0.1	-
160-170	IIB3tb	-	6.2	1.0	0.2	0.1	0.2	0.1	-
175-185	IIC+R1	-	7.4	1.0	0.3	0.1	0.1	0.1	-
200-220	III?R1+C	-	7.0	1.1	0.3	0.1	0.2	tr	-
220-240	IV?R1	-	6.8	1.0	0.4	0.1	0.2	tr	-

Table 58   Peak area percentages of X-ray diffractograms of the clay fraction, treated with Mg-ethylene glycol, and cation exchange capacity of the clay fraction after H_2O_2 pretreatment

Depth cm	Hor	d Values (Å)						CEC meq/100 g
		7	10	12	14	16	18	
0-7	A1	n.d.	79	n.d.	12.5	6.5	2	20.0
12-25	B2	n.d.	72	n.d.	18	7	3	21.3
40-50	B3	n.d.	70	n.d.	21	7	2	20.7
65-80	B3+R1	n.d.	82	n.d.	13	2	3	18.0
110-120	IIB2tb	12	85.5	n.d.	2.5	1.5	1	16.0
140-150	IIB2tb	6	93	n.d.	1.5	-	-	14.7
160-170	IIB2tb	5	94	n.d.	1.5	-	-	14.0
175-185	IIC+R1	5	94	n.d.	1.5	-	-	12.7
200-220	III?R1+C	5	94	n.d.	1.5	-	-	6.7
220-240	IV?R1	6.5	90	n.d.	2.5	1	-	4.7

n.d. = not determined

<u>Table 59</u> Pollen and charcoal content (analyst R. Slotboom)

Depth cm	Hor.	Arboreal pollen	Non-arboreal pollen	Charcoal fragments
3-6	A1	*Alnus* 3 *Pinus* 5 *Quercus* 20 *Corylus* 10 *Carpinus* 9 *Tilia* 1 *Fagus* 7 *Picea* 1 Σ AP = 56	*Fagopyrum* 3 *Umbelliferae* 1 *Rumex* 2 *Gramineae* 6 *Rosaceae* 1 *Compositae l.* 1 *Cyperaceae* 13 *Chenopod.* 1 Cerealia 14 Mon.psilatae 6 Σ NAP = 43	very much
14-16	B2	*Alnus* 8 *Corylus* 30 *Quercus* 54 *Picea* 1 *Carpinus* 14 *Betula* 6 *Fagus* 21 *Salix* 1 *Pinus* 8 Σ AP = 143	*Fagopyrum* 6 *Ericaceae* 3 *Rumex* 1 *Ranuncul.* 1 *Rosaceae* 1 *Polypodium* 2 *Cyperaceae* 24 *Botryococ.* 5 Cerealia 36 *Papilonac.* 1 *Umbelliferae* 3 *Compositae T* 17 *Gramineae* 22 *Rubiaceae* 1 *Compositae l.* 3 *Cruciferae* 2 *Chenopod.* 3 *Scrophul.* 1 Mon.psilatae 5 *Polygonacea* 2 *Sphagnum* 1 *Caryophyll.* 1 *Plantago* 5 *Lycopodium* 1 Σ NAP = 138	very much
25-30	B2	*Quercus* 4 *Corylus* 7 Σ AP = 11	*Fagopyrum* 1 Mon.psilatae 3 *Cyperaceae* 2 *Epilobium* 1 Cerealia 3 *Plantago* 1 Σ NAP = 12	very much
45-50	B3	*Alnus* 1 *Carpinus* 1 *Corylus* 6 Σ AP = 8	*Fagopyrum* 1 *Compositae l.* 4 *Cyperaceae* 4 *Plantago* 1 Cerealia 1 *Caryophyll.* 1 Mon.Psilatae 13 *Ranunculac.* 1 *Gramineae* 3 Σ NAP = 29	very much
65-70	B3+R1	Σ AP = 0	*Fagopyrum* 1 Σ NAP = 1	much
90-100	IIB2tb	Σ AP = 0	Σ NAP = 0	very little
140-150	IIB2tb	Σ AP = 0	Σ NAP = 0	very little
170-180	IIC+R1	Σ AP = 0	Σ NAP = 0	---

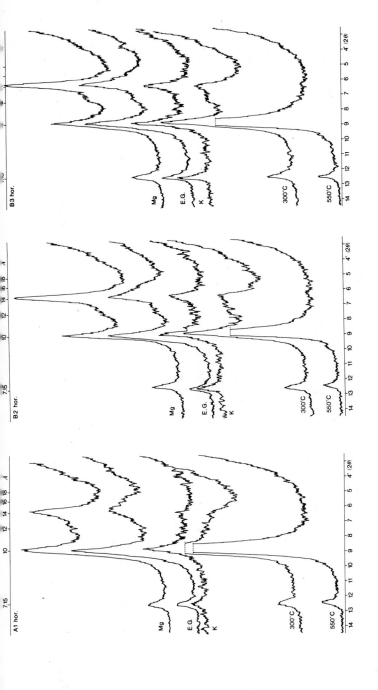

Figure 33a   X-ray diffraction patterns for oriented clay separates of profile Haarts 1
(A1, B2, B3 horizon)

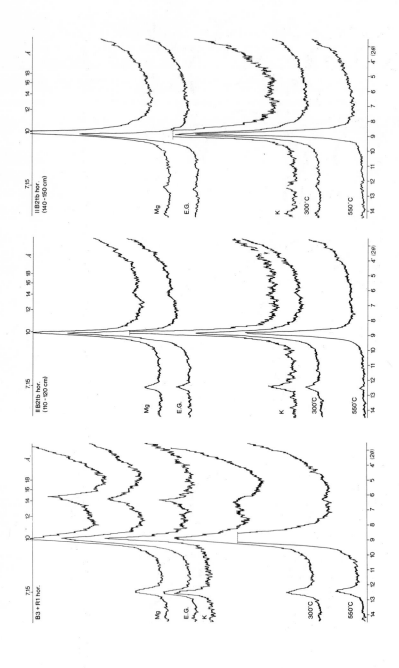

Figure 33b X-ray diffraction patterns for oriented clay separates of profile Haarts 1 (B3 + R1, IIB2tb (110-120 cm), IIB2tb (140-150 cm)

218

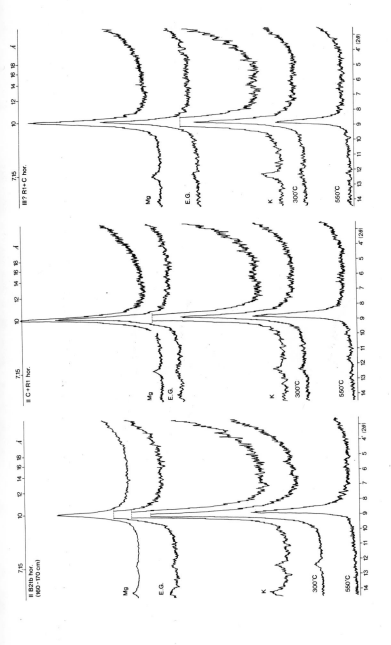

Figure 33c   X-ray diffraction patterns for oriented clay separates of profile Haarts 1 (IIBtb (160-170 cm), IIC + R1, III?R1 + C)

219

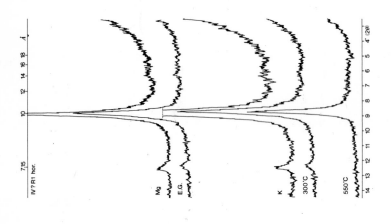

Figure 33d   X-ray diffraction patterns for oriented clay separates of profile Haarts 1 (IV?R1)

220

PROFILE SCHEISSGROND 1

Classification: Thapto-hapludalfic Umbric Dystrochrept
(dystric cambisol, overlying truncated
luvisol); loamy-skeletal, mixed, mesic

Location: see fig. 10

Elevation: 435 m

Land form: ESE facing steep lower slope

Slope: mixed oak-beech coppice

Vegetation: 23°

Profile: moist throughout; roots throughout

Profile description (abbreviated)

O   1-0 cm   very dark brown (7.5 YR 3/2) partly
decomposed organic matter; abrupt and
smooth boundary

A1   0-6 cm   dark brown (7.5 YR 3/3) gravelly loam;
weak medium crumb; very friable; clear
and smooth boundary

B1   6-30/35cm   brown (7.5 YR 4/4) gravelly stony loam;
moderate medium to fine subangular blocky;
friable; clear and smooth boundary

B2+R1 30/35-   brown to strong brown (7.5 YR 5/5)
70/80 cm   gravelly and stony loam; moderate fine to
coarse subangular blocky; friable; clear
and smooth boundary

IIB2tb 70/80-   strong brown (7.5 YR 5/6) gravelly and
110/120 cm stony loam; more stone sized quartzitic
shale fragments; moderate to strong fine
to coarse subangular blocky; friable;
gradual and smooth boundary

IIB2tb+R1   brown to light brown (7.5 YR 5/4-6/4)
+110/120cm gravelly stony loam; more stone sized
(quartzitic) shale fragments than horizon
above; moderate fine to coarse subangular
blocky; friable

Table 60 Particle size distribution, organic carbon, nitrogen, C/N ratio

Depth (cm)	Hor	Particle size					
		$(mm)^{xx}$	$(mm)^{x}$				
		> 2	2-1	1-0.5	0.5-0.25	0.25-0.1	0.1-0.5
0-6	A1	55.5	19.0	5.0	4.0	5.0	4.5
15-25	B1	60.9	18.5	5.0	4.0	5.0	4.0
45-60	B2+R1	65.4	17.5	5.0	4.5	5.5	4.5
85-100	IIB2tb	55.4	14.0	4.5	5.5	7.5	4.5
115-125	IIB2tb+R1	65.6	13.0	5.0	6.0	8.5	5.0

(continued)

Depth (cm)	Hor	$(\mu m)^{x}$		Org.C^{x}	N^{x}	C/N ratio
		50-2	< 2			
0-6	A1	41.0	21.5	3.5	0.24	14.5
15-25	B1	41.0	23.0	2.2	0.20	11.0
45-60	B2+R1	45.0	18.5	0.65	0.07	9.3
85-100	IIB2tb	52.0	12.5	0.45	0.05	9.2
115-125	IIB2tb+R1	52.5	10.5	0.3	0.04	8.0

^{x}weight perc. abs. dry fine earth; ^{xx}weight perc. air dry soil

Table 61 pH and exchangeable cation characteristics

Depth (cm)	Hor	pH		Exchangeable cations in meq/100 g fine earth			
		H_2O	$CaCl_2$	Na	K	Ca	Mg
0-6	A1	4.3	3.9	0.34	1.46	1.16	1.36
15-25	B1	4.7	3.9	0.16	0.44	0.13	0.21
45-60	B2+R1	4.8	4.0	0.15	0.36	0.14	0.80
85-100	IIB2tb	4.7	4.0	0.26	1.09	0.14	1.59
115-125	IIB2tb+R1	5.4	4.9	0.46	2.26	0.40	3.17

(continued)

Depth (cm)	Hor	Exchangeable cations in meq/100 g fine earth				Base sat.%
		Al	H	Σ=CEC	CEC^{x}	
0-6	A1	8.61	2.21	15.1		28.5
15-25	B1	6.10	1.00	8.0		11.7
45-60	B2+R1	4.36	1.06	6.9	7.6	21.1
85-100	IIB2tb	3.13	0.83	7.0	6.8	43.8
115-125	IIB2tb+R1	2.18	0.69	9.2	8.9	68.7

^{x}CEC -NH_4 acetate (1.0 N, pH = 7.0)

222

Table 62 Extractable iron, aluminium and manganese by dithionite, ammonium oxalate and sodium pyrophosphate in the fine earth fraction (weight percentages)[x]

Depth (cm)	Hor	Fe_d	Fe_o	Fe_p	Fe_{t-d}	Fe_o/Fe_d	Fe_p/Fe_o
0-6	A1	2.47	0.56	0.31	1.73	0.23	0.55
15-25	B1	2.46	0.60	0.20	1.94	0.24	0.33
45-60	B2+R1	2.24	0.50	0.11	1.74	0.22	0.22
85-100	IIB2tb	1.75	0.33	0.06	1.86	0.19	0.19
115-125	IIB2tb+R1	1.81	0.17	0.03	1.78	0.09	0.20

(continued)

Depth (cm)	Hor	Al_d	Al_o	Al_p	Mn_d	Mn_o	Mn_p	Mn_{t-d}
0-6	A1	0.38	0.36	0.21	0.21	0.21	0.11	0
15-25	B1	0.41	0.38	0.19	0.22	0.15	0.05	0
45-60	B2+R1	0.29	0.28	0.12	0.09	0.06	0.02	0.03
85-100	IIB2tb	0.21	0.19	0.08	0.06	0.05	0.01	0.03
115-125	IIB2tb+R1	0.19	0.15	0.06	0.09	0.06	0.01	0.03

[x] organic matter free and abs. dry base

Table 63 Elemental composition of the Li-saturated clay fraction (weight perc.[x]) and derived molar ratios

Depth (cm)	Hor	SiO_2	Al_2O_3	Fe_2O_3	FeO	MnO	MgO	CaO	Na_2O	K_2O
0-6	A1	46.2	26.3	8.84	0.96	0.10	1.54	0.13	0.25	4.03
15-25	B1	44.2	28.1	9.62	0.87	0.27	1.62	0.12	0.30	4.58
45-60	B2+R1	44.0	28.3	9.82	0.80	0.20	1.73	0.02	0.27	4.15
85-100	IIB2tb	45.4	26.6	10.0	0.87	0.20	1.79	0.02	0.26	3.96
115-125	IIB2tb+R1	46.5	25.5	10.0	0.91	0.22	1.81	0.03	0.25	4.19

Table 63 - *(continued)*

Depth (cm)	Hor	TiO_2	P_2O_5	Li_2O	LOI	SiO_2/R_2O_3	SiO_2/Al_2O_3	SiO_2/Fe_2O_3	Al_2O_3/Fe_2O_3
0-6	A1	0.79	0.81	0.10	9.95	2.40	2.98	12.40	4.16
15-25	B1	0.77	0.68	0.10	8.64	2.15	2.67	11.10	4.16
45-60	B2+R1	0.85	0.73	0.10	9.08	2.13	2.64	10.92	4.14
85-100	IIB2tb	0.91	0.56	0.09	9.42	2.29	2.90	10.99	3.79
115-125	IIB2tb+R1	1.07	0.45	0.14	8.94	2.42	3.10	11.20	3.62

^{x} abs. dry base

Table 64 Normative mineralogical composition of the clay fraction (weight perc.)

Depth (cm)	Hor	Q	Ab	Ill	Kaol	Chlor	Fe-Chlor
0-6	A1	9.9	2.1	49	20	-	-
15-25	B1	6.8	2.5	52	19	-	-
45-60	B2+R1	5.0	2.3	50	23	-	-
85-100	IIB2tb	8.4	2.3	48	21	-	-
		(12.4)	(2.3)	(48)	(21)	(6.3)	(2.1)
115-125	IIB2tb+R1	10.7	2.1	50	16	-	-
		(14.8)	(2.1)	(50)	(11)	(6.4)	(2.2)

(continued)

Depth (cm)	Hor	Verm	Fe-Verm	Go	Apa	Str	MnO_2	Ru
0-6	A1	5.1	2.2	9.1	0.2	1.8	0.1	0.8
15-25	B1	5.2	2.0	10.1	0.2	1.5	0.3	0.8
45-60	B2+R1	5.6	1.8	10.1	0.1	1.8	0.2	0.9
85-100	IIB2tb	5.8	2.0	10.6	0.1	1.4	0.3	0.9
		(-)	(-)	(10.7)	(0.1)	(1.4)	(0.3)	(0.9)
115-125	IIB2tb+R1	5.9	2.0	10.8	0.1	1.2	0.3	1.0
		(-)	(-)	(10.9)	(0.1)	(1.2)	(0.3)	(1.0)

Table 65 Elemental composition of the fine earth fraction (weight perc.[x]) and derived molar ratios

Depth (cm)	Hor	SiO_2	Al_2O_3	Fe_2O_3	FeO	MnO	MgO	CaO	Na_2O	K_2O
0-6	A1	59.5	12.9	4.79	0.74	0.24	0.84	0.03	0.40	2.44
15-25	B1	66.1	14.8	5.24	0.73	0.25	0.96	0.01	0.47	2.85
45-60	B2+R1	69.4	14.3	4.95	0.61	0.15	0.95	0.01	0.54	2.71
85-100	IIB2tb	72.5	12.8	4.38	0.67	0.12	0.92	0.01	0.58	2.74
115-125	IIB2tb+R1	73.3	12.4	4.34	0.68	0.14	0.91	0.01	0.61	2.68

(continued)

Depth (cm)	Hor	TiO_2	P_2O_5	SO_3	LOI	SiO_2/R_2O_3	SiO_2/Al_2O_3	SiO_2/Fe_2O_3	Al_2O_3/Fe_2O_3
0-6	A1	0.91	0.21	0.07	16.9	6.12	7.82	28.19	3.60
15-25	B1	1.01	0.17	0.08	7.38	6.02	7.60	29.04	3.82
45-60	B2+R1	1.00	0.13	0.05	5.20	6.58	8.24	32.79	3.98
85-100	IIB2tb	0.94	0.11	0.04	4.23	7.64	9.59	37.59	3.92
115-125	IIB2tb+R1	0.94	0.11	0.05	3.84	7.96	10.05	38.25	3.81

[x] abs. dry base

Table 66 Normative mineralogical composition of the fine earth fraction (weight perc.)

Depth (cm)	Hor	Q	Ab	K-mica	Kaol	Chlor	Fe-Chlor	Verm
0-6	A1	50	3.9	27	6.7	2.2	1.5	1.1
15-25	B1	49	4.1	28	6.2	2.2	1.6	1.2
45-60	B2+R1	51	4.7	26	6.6	2.2	1.2	1.0
85-100	IIB2tb	56 (57)	5.0 (5.0)	25 (25)	2.9 (2.3)	2.5 (3.3)	1.4 (1.6)	0.7 (-)
115-125	IIB2tb+R1	58 (58)	5.2 (5.2)	24 (24)	2.1 (1.6)	2.6 (3.2)	1.4 (1.6)	0.6 (-)

Table 66 - *(continued)*

Depth (cm)	Hor	Fe-Verm	Hm+Go	Ru	MnO_2	Str	Apa	Pr
0-6	A1	0.5	5.5	1.0	0.3	0.6	0.04	0.1
15-25	B1	0.5	5.5	1.0	0.3	0.4	0.05	0.1
45-60	B2+R1	0.3	5.0	1.0	0.2	0.4	0.02	0.1
85-100	IIB2tb	0.3 (-)	4.5 (4.5)	0.9 (0.9)	0.2 (0.2)	0.3 (0.3)	0.01 (0.01)	0.1 (0.1)
115-125	IIB2tb+R1	0.2 (-)	4.4 (4.4)	0.9 (0.9)	0.2 (0.2)	0.3 (0.3)	0.01 (0.01)	0.1 (0.1)

Table 67 Peak area percentage of X-ray diffractograms of the clay fraction, treated with Mg-ethylene glycol, and cation exchange capacity of the clay fractions after H_2O_2 treatment

Depth (cm)	Hor	d values (Å)						CEC meq/100 g
		7	10	12	14	16	17-18	
0-6	A1	n.d.	80	n.d.	12	6	2	7
15-25	B1	n.d.	76	n.d.	15	6	3	7
45-60	B2	n.d.	64	n.d.	22.5	11	2.5	8
85-100	IIB2tb	n.d.	72.5	n.d.	17	7	3.5	6
115-125	IIB2tb+R1	n.d.	73	n.d.	13	9	5	9

n.d. = not determined

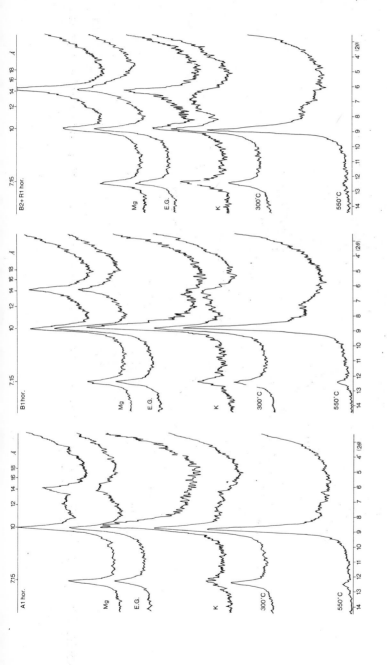

Figure 34a X-ray diffraction patterns for oriented clay separates of profile Scheissgrond 1
(A1, B1, B2 + R1)

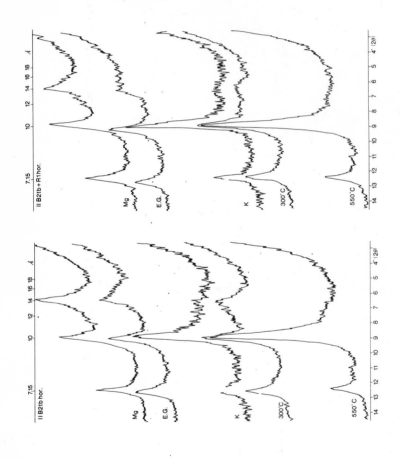

Figure 34b   X-ray diffraction patterns for oriented clay separates of profile Scheissgrond 1 (IIB2tb, IIB2tb + R1)

PROFILE HAARTS 3 (from Kwaad & Mücher 1976)

Classification: Mollic Hapludalf (orthic acrisol); loamy skeletal, mixed, mesic

Location: 250 m NNE of Haarts catchment (see fig. 10)

Elevation: 430 m

Land form: upper dry valley

Slope: 3°

Vegetation: mixed oak-beech forest

Profile: moist throughout; roots throughout

Profile description (abbreviated)

A1   0-2 cm   dark brown (7.5 YR 3/2) slightly gravelly loam; medium crumb; loose; abrupt and smooth boundary

C   2-47 cm   dark reddish brown (5 YR 3/4) slightly gravelly loam; fine crumbs (inherited); loose; abrupt and irregular boundary

IIB2tb 47-85 cm   yellowish brown (10 YR 5/6) to brownish yellow (10 YR 6/6) very gravelly to stony loam; many to very many fresh quartzite sandstone and shale fragments, mostly parallel to the soil surface; subangular blocky; firm; patchy thin ferriargillans; gradual and wavy boundary

IIB2tb + R1   yellowish red (5 YR 5/8) and pale brown
    85-100+cm   (10 YR 6/3) very stony loam; very many quartzitic shale and shale fragments, standing upright; patchy thin ferriargillans

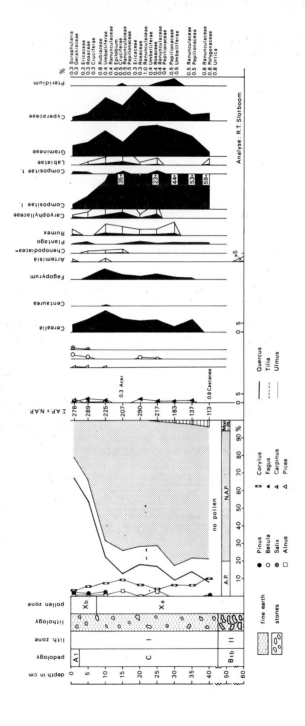

Figure 35   Pollen diagram of profile Haarts 3

Table 68 Chemical composition of the rocks in the Haarts catchment

sample no.	SiO$_2$	Al$_2$O$_3$	Fe$_2$O$_3$	FeO	MnO	MgO	CaO	Na$_2$O	K$_2$O	TiO$_2$	P$_2$O$_5$	CO$_2$	SO$_3$	LOI*
non weathered rocks (grayish coloured)														
41R	71.5	10.8	1.68	3.93	0.24	1.70	0.15	1.12	2.16	0.67	0.11	3.61	0.06	5.97
179R	64.5	17.8	2.10	3.10	0.11	2.20	0.20	1.08	3.66	0.94	0.14	0.23	0.10	3.59
180R	57.1	22.7	2.62	3.12	0.09	2.05	0.23	0.66	5.46	1.10	0.14	0.26	0.11	4.50
184R	67.7	15.3	3.23	2.36	0.10	1.58	0.26	1.19	3.26	1.00	0.16	0.15	0.10	3.59
185R	68.6	15.2	2.70	2.52	0.10	1.64	0.21	1.22	3.19	0.99	0.16	0.18	0.12	3.34
very slightly weathered rocks (light brown coloured)														
189R	66.5	17.2	4.02	1.82	0.14	1.68	0.08	0.48	3.24	0.84	0.10	0.12	0.08	3.67
190R	81.6	8.59	2.40	1.44	0.29	1.04	0.11	0.64	1.37	0.45	0.07	0.10	0.07	1.83
191R	80.2	0.60	4.11	0.86	0.15	0.70	0.09	0.82	1.51	0.52	0.31	-	0.06	1.25
slightly weathered rocks (brown coloured)														
215R	63.0	18.4	7.77	0.12	0.10	0.47	0.03	1.02	3.55	0.98	0.12	-	0.19	4.14
216R	67.3	17.8	4.61	0.11	0.16	0.45	0.05	1.12	3.34	0.91	0.10	-	0.20	3.81
42R	68.3	12.1	10.7	0.05	0.38	0.31	0.01	1.09	2.29	0.76	0.16	0.23	0.02	3.87
43R	70.0	10.9	10.5	0.05	0.39	0.30	0.01	1.09	2.10	0.71	0.15	0.29	0.03	3.93
182R	73.7	10.2	8.04	0.38	0.19	0.63	0.03	0.96	1.98	0.64	0.12	-	0.06	3.07
183R	73.9	7.94	10.9	0.17	0.31	0.46	0.03	0.94	1.41	0.52	0.11	-	0.07	3.28
186R	72.3	14.1	5.02	0.10	0.20	0.28	0.15	1.09	2.71	0.84	0.16	-	0.06	3.13
187R	79.8	7.07	7.04	0.04	0.35	0.20	0.14	0.69	1.25	0.67	0.16	-	0.04	2.55
188R	68.2	17.1	5.46	0.07	0.19	0.34	0.19	1.19	3.08	0.84	0.14	-	0.06	3.21
192R	80.4	8.08	6.24	0.05	0.33	0.15	0.11	0.67	1.21	0.75	0.15	-	0.05	1.94

* LOI = loss of ignition

Table 69 Normative mineralogical composition (weight perc.) of the rocks in the Haarts catchment

Sample no.	Q	Ab	Ms	Kaol	Chlor	Hm	Ru	Sid	Apa	Str	Pr	Zo	MnO$_2$
non-weathered rocks (grayish coloured)													
41R	56.5	9.6	18.5	-	5.6	1.7	0.7	7.0	0.3	-	0.1	-	-
179R	41.8	9.1	31.3	-	13.8	2.0	0.9	0.6	0.3	-	0.1	-	-
180R	29.7	5.6	46.7	-	13.1	2.6	1.1	0.7	0.3	-	0.1	0.1	-
184R	46.5	10.2	27.7	-	10.0	3.2	1.1	0.4	0.4	-	0.1	0.2	-
185R	47.4	10.4	27.1	-	10.4	2.7	0.9	0.5	0.4	-	0.1	-	-
very slightly weathered hard rocks (grayish brown coloured)													
189R	48.1	4.2	28.2	3.5	10.2	4.1	0.8	0.3	0.1	0.2	0.1	-	-
190R	70.6	5.4	11.5	1.7	7.1	2.4	0.4	0.3	0.3	-	0.1	0.1	-
191R	66.2	6.8	12.6	4.7	4.2	3.8	0.5	0.4	0.2	0.5	0.1	-	-
173R	66.9	3.0	15.1	7.4	3.0	3.7	0.5	tr	0.1	0.2	0.1	-	-
slightly weathered hard rocks (brown coloured)													
215R	38.1	8.7	30.1	11.8	1.9	7.7	0.9	-	0.1	0.3	0.2	-	0.1
216R	42.7	9.5	28.4	11.4	1.9	4.5	0.9	-	0.1	0.2	0.2	-	0.1
42R	51.0	9.3	19.6	6.4	1.1	10.6	0.8	0.2	-	0.5	0.1	-	0.4
43R	54.2	9.3	18.1	5.0	1.1	10.5	0.8	0.2	-	0.3	0.1	-	0.4
182R	59.2	8.3	16.9	2.9	3.2	8.1	0.7	-	0.1	0.3	0.1	-	0.2
183R	62.6	8.2	12.2	3.0	1.7	11.0	0.5	-	0.1	0.3	0.1	-	0.4
186R	52.1	9.4	23.1	7.8	1.2	4.9	0.8	-	0.3	0.2	0.1	-	0.2
187R	70.0	5.9	10.7	4.0	0.8	7.1	0.7	-	0.2	0.2	0.1	-	0.4
188R	43.8	10.1	26.0	11.8	1.4	5.5	0.8	-	0.3	-	0.1	-	0.2
192R	68.6	5.6	10.2	7.1	0.7	6.2	0.8	-	0.2	0.2	0.1	-	0.4

Table 70 X-ray diffraction of the main constituents of the powdered rock

Sample no.	Chlorite	K-mica	Albite	Siderite	Quartz
	non-weathered rocks (grayish coloured)				
41R	tr	+++(+)		xx	50-80
179R	++	+++	xx	-	50-80
180R	++	+++	x(x)	-	30-50
184R	+(+)	++	xx(x)	-	60-90
185R	+	+(+)	xx	-	60-90
	very slightly weathered hard rocks (grayish brown coloured)				
189R	+(+)	++	x	-	50-80
190R	+(+)	+	x	-	60-90
191R	(+)	+	x	-	60-90
173R	(+)	+	(x)	-	60-90
	slightly weathered hard rocks (brown coloured)				
215R	tr	++(+)	(x)	-	15-25
216R	tr	+++	x	-	25-40
42R	?	++(+)	x	?	40-60
43R	?	++(+)	x	?	40-60
182R	(+)	+	x	-	60-90
183R	(+)	+	x	-	60-90
186R	?	+	(x)	-	50-80
187R	?	+	(x)	-	60-90
188R	?	+(+)	x	-	50-80
192R	tr	(+)	(x)	-	60-90

Table 71   Chemical composition of the springwaters in the Haarts catchment
(<< = << 0.001)

SPRING no.	DATE	pH	K⁺	Na⁺	NH₄⁺	Ca²⁺	Mg²⁺	H₄SiO₄	NO₃⁻	Cl⁻	HCO₃⁻ (Alk)	SO₄²⁻	EC₂₅ µS/cm	Temp. °C	Σcat meq/l	Σan meq/l
						mmoles/l										
1	11.5.73	6.35	0.011	0.174		0.101	0.137	0.113	0.075	0.149	0.309	0.062	72		0.72	0.66
	19.5.73	6.69	0.016	0.177		0.108	0.132	0.105	0.073	0.148	0.340	0.046	70		0.72	0.65
	25.5.73	6.51	0.015	0.182		0.076	0.166	0.116	0.014	0.148	0.301	0.071	68		0.71	0.61
	29.5.73	6.51	0.013	0.152		0.089	0.140	0.129	0.057	0.173	0.301	0.029	70		0.64	0.59
	7.7.73	7.10	0.027	0.182		0.151	0.259		0.054	0.141	0.755	0.067	108		1.00	1.08
	15.8.73	7.00	0.029	0.217		0.174	0.340	0.126	0.053	0.107	1.198	0.076	124	12.5	1.40	1.51
	12.9.73	7.29	0.041	0.240		0.226	0.404	0.132	0.074	0.163	1.200	0.076	145	11.8	1.54	1.59
	10.10.73	7.08	0.032	0.250	0.043	0.233	0.428	0.127	0.028	0.155	1.356	0.067	155	10.2	1.65	1.67
	6.11.73	6.50	0.032	0.223		0.204	0.408	0.132			1.190	0.083	135	6.2	1.48	
	12.12.73	6.45	0.015	0.169	0.001	0.296	0.160	0.088	0.183	0.174	0.340	0.063	81		1.10	0.82
	8.1.74	6.20	0.016	0.193		0.092	0.166	0.110	0.139	0.207	0.265	0.089	79	6.2	0.73	0.79
	12.2.74	6.11	0.016	0.181		0.091	0.154	0.098	0.216	0.256	0.105	0.076	82		0.68	0.65
	19.3.74	6.06	0.014	0.167		0.084	0.160	0.107	0.134	0.213	0.200	0.074	77	6.6	0.67	0.69
	23.4.74	6.80	0.016	0.185		0.100	0.183	0.121	0.085	0.205	0.425	0.068	90	9.0	0.77	0.85
	30.5.74	7.11	0.013	0.158		0.114	0.213	0.125	0.070	0.205	0.517	0.061	91	8.6	0.83	0.91
	4.7.74	7.18	0.020	0.242		0.142	0.259	0.130	0.064	0.177	0.726	0.070	106	8.9	1.06	1.10
	8.8.74	7.40	0.022	0.261		0.156	0.278	0.127	0.074	0.183	0.807	0.068	114	10.8	1.15	1.20
	16.9.74	7.40	0.024	0.234		0.193	0.343	0.129	0.060	0.132	0.988	0.057	128	12.0	1.33	1.30
	23.10.74	6.29	0.010	0.136	<<	0.101	0.117	0.152	0.192	0.265	0.203	0.035	87	7.5	0.58	0.73
	21.11.74	6.50	0.018	0.197	<<	0.101	0.164	0.118	0.166	0.229	0.215	0.080	82	7.8	0.76	0.81
	17.12.74	6.15	0.011	0.171	<<	0.108	0.142	0.117	0.179	0.244	0.164	0.102	91	7.4	0.68	0.79
	22.1.75	6.45	0.16	0.211	<<	0.108	0.149	0.124	0.153	0.252	0.205	0.047	81	6.8	0.74	0.70
	19.2.75	6.68	0.016	0.184	<<	0.108	0.174	0.114	0.172	0.212	0.248	0.092	91	6.6	0.78	0.78
	13.5.75	6.48	0.009	0.155	<<	0.072	0.125	0.123	0.139	0.219	0.290	0.064	84	7.2	0.55	0.78
	15.5.75	6.58	0.015	0.211	<<	0.100	0.178	0.126	0.139	0.227	0.315	0.064	84	6.8	0.77	0.81
	27.5.75	6.79	0.016	0.216	<<	0.105	0.193	0.135	0.119	0.210	0.366	0.057	87	7.8	0.83	0.82
	15.7.75	7.09	0.021	0.241	<<	0.145	0.262	0.131	0.073	0.197	0.675	0.036	107	12.0	1.08	1.02
	27.8.75	7.20	0.021	0.253	<<	0.163	0.301	0.130	0.089	0.176	0.791	0.057	120	11.2	1.20	1.17
	15.10.75	7.19	0.025	0.253	<<	0.174	0.324	0.133	0.096	0.195	0.846	0.050	127	8.2	1.27	1.24

Table 72   Chemical composition of the springwaters in the Haarts catchment (<< = << 0.001)

SPRING no.	DATE	pH	K⁺	Na⁺	NH₄⁺ (mmoles/l)	Ca²⁺	Mg²⁺	H₄SiO₄	NO₃⁻	Cl⁻	HCO₃⁻ (Alk)	SO₄²⁻	EC₂₅ µs/cm	Temp °C	Σcat (meq/l)	Σan (meq/l)
2	11.5.73	6.30	0.010	0.196		0.090	0.206	0.118	0.062	0.176	0.455	0.058	81		0.84	0.81
	19.5.73	6.50	0.015	0.180		0.126	0.169	0.109	0.047	0.165	0.515	0.042	79		0.82	0.81
	25.5.73	6.30	0.011	0.196		0.092	0.185	0.124	0.029	0.151	0.418	0.063	78		0.79	0.74
	29.5.73	6.41	0.015	0.174		0.092	0.189	0.123	0.047	0.192	0.439	0.042	79		0.77	0.76
	15.8.73	6.60	0.024	0.239		0.195	0.459	0.133	0.030	0.136	1.270	0.055	149	9.5	1.68	1.69
	12.9.73	6.65	0.025	0.256		0.212	0.491	0.142	0.045	0.157	1.420	0.052	159	9.5	1.70	1.73
	10.10.73	6.60	0.022	0.246	<<	0.224	0.505	0.137	0.020	0.126	1.400	0.055	163	9.4	1.74	1.69
	5.11.73	6.40	0.032	0.246	<<	0.187	0.444	0.120	0.025	0.141	1.161	0.085	142	8.7	1.54	1.50
	12.12.73	6.41	0.012	0.160	<<	0.143	0.165	0.125	0.140	0.189	0.353	0.058	92		0.79	0.80
	8.1.74	6.25	0.016	0.204	0.001	0.132	0.240	0.117	0.124	0.199	0.400	0.083	95	7.3	0.96	0.89
	12.2.74	6.06	0.012	0.193	0.013	0.095	0.164	0.112	0.198	0.230	0.225	0.074	83		0.74	0.80
	19.3.74	6.11	0.010	0.152	<<	0.082	0.141	0.114	0.117	0.205	0.319	0.092	83	6.4	0.61	0.82
	23.4.74	6.48	0.009	0.181	<<	0.095	0.177	0.122	0.081	0.229	0.654	0.077	104	7.2	0.73	1.12
	30.5.74	6.60	0.016	0.204	0.001	0.137	0.259	0.130	0.063	0.193	0.720	0.071	105	7.5	1.01	1.12
	4.7.74	6.70	0.017	0.226	0.001	0.160	0.320	0.129	0.055	0.187	0.937	0.073	124	8.2	1.21	1.32
	8.8.74	6.67	0.013	0.235	0.001	0.143	0.382	0.137	0.053	0.191	1.130	0.079	137	8.8	1.05	1.53
	16.9.74	6.75	0.026	0.260	<<	0.213	0.447	0.138	0.043	0.136	1.336	0.072	157	9.6	1.63	1.66
	23.10.74	6.19	0.014	0.182		0.097	0.169	0.126	0.173	0.259	0.320	0.044	95	7.5	0.73	0.84
	21.11.74	6.38	0.011	0.160	<<	0.077	0.137	0.124	0.126	0.227	0.380	0.078	88	7.8	0.60	0.87
	17.12.74	6.19	0.015	0.211		0.104	0.183	0.143	0.185	0.234	0.233	0.079	87	7.4	0.80	0.81
	22.1.75	6.35	0.013	0.217		0.100	0.181	0.124	0.126	0.206	0.304	0.073	86	6.5	0.79	0.78
	22.1.75	6.35	0.013	0.217		0.102	0.190	0.123	0.126	0.237	0.317	0.080	86	7.1	0.81	0.84
	23.1.75	6.35	0.011	0.196		0.094	0.169	0.122	0.126	0.204	0.320	0.082	85		0.73	0.82

Table 72 (continued)

Date				<<											
8.2.75	**6.44**	0.015	**0.227**		0.116	0.218	**0.123**	**0.117**	**0.220**	0.434	0.075	95	7.0	0.91	0.92
19.2.75	6.41	0.014	0.197		0.116	0.216	**0.126**	0.126	**0.211**	0.450	0.072	94	7.1	0.88	0.92
20.2.75	6.40	0.015	0.196		0.111	0.211	0.127	0.111	0.230	0.450	0.091	93		0.86	0.97
18.3.75	6.51	0.014	0.190		0.104	0.209	0.121	0.100	0.236	0.457	0.094	92	6.7	0.83	0.79
19.3.75	6.50	0.014	0.187		0.109	0.211	0.124	0.106	0.244	0.491	0.091	90	6.6	0.84	0.97
12.5.75	6.38	0.014	0.187		0.102	0.197	0.130	0.106	0.231	0.418	0.088	110	6.8	0.79	0.91
13.5.75	6.42	0.013	0.187		0.115	0.195	0.131	0.079	0.229	0.415	0.096	88	6.8	0.82	0.92
15.5.75	6.39	0.014	0.185		0.110	0.201	0.133	**0.089**	0.182	0.429	0.089	89	6.8	0.82	0.86
26.5.75	6.60	0.015	0.196		0.113	0.229	0.133	**0.087**	0.207	0.546	0.084	96	7.2	0.90	0.99
27.5.75	6.60	0.015	0.187		0.116	0.231	0.133	0.083	0.183	0.580	0.090	96	7.1	0.90	1.00
23.6.75	6.60	0.017	0.196		0.151	0.286	0.133	**0.078**	0.184	0.734	0.064			1.09	1.11
14.7.75	6.71	0.021	0.242		0.175	0.356	0.134	0.063	0.157	0.961	0.109	125	8.2	1.33	1.40
15.7.75	6.67	0.021	0.242		0.175	0.358	0.134	0.060	0.141	0.969	0.111	126	8.2	1.33	1.40
26.8.75	6.60	0.023	0.264		0.207	0.435	0.137	**0.064**	0.148	1.184	0.106	147	9.5	1.57	1.61
27.8.75	6.60	0.023	0.264		0.197	0.445	0.138	0.073	0.151	1.190	0.117	149	9.5	1.57	1.65
17.9.75	6.64	0.024	0.269		0.208	0.467	0.139	0.050	0.153	1.310	0.107	157		1.64	1.73
14.10.75	6.70	0.024	0.269		0.218	0.461	0.140	0.050	0.153	1.235	0.106	152	8.9	1.65	1.65
15.10.75	6.68	0.024	0.269		0.215	0.452	0.139	0.060	0.146	1.242	0.120	152	9.0	1.63	1.69
26.10.75	6.56	0.025	0.269		0.223	0.461	0.139	0.053	0.144	1.265	0.098	154		1.66	1.66
13.4.76	6.57	0.019	0.226		0.133	0.269	0.130	0.089	0.202	0.620	0.057	104		1.05	1.02
14.4.76	6.55	0.017	0.226		0.132	0.279	0.128	0.087	0.208	0.620	0.068	105		1.06	1.05

237

Table 73   Chemical composition of the springwaters in the Haarts catchment (<< = << 0.001)

SPRING no.	DATE	pH	K+	Na+	NH4+	Ca2+	Mg2+	H4SiO4	NO3-	Cl-	HCO3- (Alk)	SO4 2-	EC25 µS/cm	Temp °C	Σcat meq/l	Σan meq/l
					mmoles/l											
3	27.5.73	6.39	0.009	0.178		0.095	0.207	0.124	0.022	0.137	0.441	0.066	79		0.85	0.73
	29.5.73	6.35	0.014	0.153		0.093	0.222	0.128	0.030	0.173	0.459	0.060	80		0.81	0.78
	6.11.73	6.40	0.025	0.259		0.211	0.440	0.100	0.036	0.144	1.323	0.081	157	7.9	1.67	1.67
	12.12.73	6.50	0.018	0.224		0.158	0.328	0.128	0.068	0.233	0.767	0.078	126		1.21	1.22
	8.1.74	6.42	0.017	0.204		0.137	0.315	0.132	0.009	0.153	0.857	0.039	107	6.9	1.13	1.10
	13.2.74	6.45	0.010	0.158		0.071	0.157	0.120	0.003	0.191	0.302	0.031	67	7.5	0.62	0.56
	19.3.74	6.40	0.014	0.163	<<	0.097	0.223	0.132	<<		0.495	0.050	85	6.8	0.82	
	23.5.74	6.51	0.035	0.212	0.001	0.141	0.317	0.150	<<	0.119	0.844	0.039	112	7.2	1.16	1.04
	30.5.74	6.71	0.016	0.209	<<	0.136	0.296	0.150	0.003	0.117	0.866	0.065	111	7.7	1.09	1.12
	4.7.74	6.66	0.018	0.244	<<	0.177	0.389	0.147	0.036	0.136	1.095	0.067	133	8.8	1.39	1.40
	8.8.74	6.71	0.022	0.261		0.217	0.504	0.151	0.007	0.100	1.492	0.057	157	9.9	1.73	1.71
	16.9.74	6.75	0.025	0.272		0.246	0.509	0.151	0.005	0.100	1.690	0.048	184	11.0	1.81	1.89
	23.10.74	6.69	0.017	0.190		0.129	0.293	0.141	0.005	0.108	0.770	0.083	108	7.5	1.05	1.05
	21.11.74	6.43	0.016	0.193	<<	0.103	0.206	0.149	0.004	0.185	0.494	0.087	81	7.9	0.83	0.85
	17.12.74	6.31	0.010	0.171		0.068	0.158	0.145		0.183	0.323	0.056	66	7.4	0.63	0.62
	18.12.74	6.31	0.010	0.169		0.060	0.131	0.138	0.003	0.181	0.274	0.071	62	7.8	0.57	0.60
	22.1.75	6.35	0.011	0.159		0.081	0.161	0.146	0.007	0.166	0.476	0.075	76	7.5	0.65	0.79
	19.2.75	6.65	0.012	0.170	<<	0.105	0.234	0.148	0.006	0.149	0.618	0.090	87	6.9	0.86	0.95
	19.3.75	6.65	0.011	0.198		0.111	0.240	0.150	0.009	0.190	0.597	0.087	85	6.8	0.91	0.97
	13.5.75	6.30	0.015	0.181		0.093	0.200	0.150	0.009	0.149	0.495	0.100	72.5	7.5	0.78	0.85
	15.5.75	6.51	0.015	0.190		0.094	0.200	0.151	0.008	0.153	0.505	0.091	74.5	7.0	0.79	0.85
	27.5.75	6.59	0.013	0.209		0.116	0.254	0.153	0.009	0.144	0.681	0.069	92.5	7.4	0.96	0.97
	15.7.75	6.78	0.019	0.269		0.210	0.434	0.152	0.003	0.164	1.358	0.095	145	8.5	1.58	1.71
	27.8.75	6.41	0.020	0.258		0.192	0.430	0.147	0.012	0.202	1.182	0.079	138	11.0	1.52	1.55
	15.10.75	6.88	0.023	0.291		0.252	0.528	0.150	0.001	0.189	1.622	0.079	173		1.87	1.97

Table 74 Chemical composition of the springwaters in the Haarts catchment (<< = << 0.001)

SPRING no.	DATE	pH	K$^+$	Na$^+$	NH$_4^+$	Ca^{2+}	Mg^{2+}	H$_4$SiO$_4$	NO$_3^-$	Cl$^-$	HCO$_3^-$ (Alk)	SO$_4^{2-}$	EC$_{25}$ µS/cm	Temp °C	Σcat meq/l	Σan meq/l
								mmoles/l								
4	13.2.74	6.20	0.014	0.177		0.087	0.174	0.123	0.077	0.195	0.285	0.039	76		0.71	0.64
	19.3.74	6.15	0.012	0.177	<<	0.087	0.179	0.114	0.032	0.170	0.331	0.042	76	6.8	0.72	0.62
	22.4.74	6.50	0.016	0.185	<<	0.125	0.245	0.133	0.044	0.187	0.623	0.052	97	7.6	0.94	0.96
	30.5.74	6.46	0.015	0.209	<<	0.124	0.238	0.135	0.031	0.153	0.655	0.056	98	7.9	0.95	0.95
	4.7.74	6.61	0.018	0.254	<<	0.155	0.312	0.134	0.041	0.153	0.891	0.058	118	8.9	1.21	1.20
	8.8.74	6.60	0.021	0.239		0.179	0.331	0.140	0.026	0.138	1.014	0.053	126	10.0	1.28	1.28
	16.9.74	6.60	0.023	0.242	<<	0.185	0.331	0.135	0.008	0.164	1.136	0.075	139	10.8	1.30	1.46
	23.10.74	6.31	0.013	0.209		0.102	0.230	0.129	0.100	0.204	0.416	0.052	92	8.2	0.88	0.83
	21.11.74	6.43	0.014	0.198	<<	0.103	0.195	0.136	0.054	0.202	0.500	0.074	84	8.1	0.84	0.89
	17.12.74	6.38	0.011	0.174	<<	0.085	0.154	0.119	0.074	0.210	0.283	0.075	75	7.3	0.66	0.72
	22.1.75	6.40	0.013	0.194		0.089	0.154	0.129	0.062	0.201	0.366	0.079	76	6.7	0.70	0.79
	19.2.75	6.32	0.016	0.205	<<	0.105	0.200	0.129	0.071	0.186	0.472	0.087	87	6.6	0.82	0.90
	19.3.75	6.50	0.013	0.205		0.100	0.186	0.129	0.054	0.186	0.466	0.062	83	7.1	0.79	0.83
	13.5.75	6.40	0.015	0.205		0.096	0.172	0.132	0.041	0.173	0.455	0.067	78	6.8	0.76	0.80
	15.5.75	6.38	0.015	0.192		0.096	0.203	0.136	0.041	0.188	0.458	0.077	80	7.4	0.81	0.84
	27.5.75	6.52	0.015	0.212		0.111	0.225	0.135	0.045	0.193	0.551	0.079	89	8.2	0.90	0.95
	15.7.75	6.51	0.020	0.243		0.159	0.318	0.139	0.037	0.190	0.845	0.079	115	11.0	1.22	1.23
	27.8.75	6.59	0.021	0.238		0.183	0.360	0.142	0.039	0.217	0.967	0.092	126	9.5	1.37	1.41
	15.10.75	6.59	0.021	0.261		0.190	0.376	0.132	0.055	0.195	1.008	0.104	129		1.39	1.47

Table 75   Chemical composition of the soil waters in the Haarts catchment (porous cups)
(< = < 0.001)

Horizon	depth cm	Date	pH	K^+	Na^+	NH_4^+	Ca^{2+}	Mg^{2+}	Mn^{2+}	Al^{3+}	H_4SiO_4	NO_3^-	Cl^-	Alk	SO_4^{2-}	EC_{25} µS/cm	Σ_{cat} meq/1	Σ_{an} meq/1
							mmoles/1									µS/cm	meq/1	
A1	10	20.3.75	5.20	0.068	0.120		0.059	0.078		0.0084	0.112		0.163	0.06		68		
		12.5.75	5.30	0.075	0.115		0.070	0.104		0.0116	0.115		0.120	0.08		70		
		15.5.75	5.29	0.077	0.130		0.068	0.083		0.0096	0.113		0.130	0.06		63		
		26.5.75	5.64	0.098	0.140		0.074	0.088		0.0079	0.116		0.109	0.17		80		
B2	35	19.11.74	5.11	0.024	0.114		0.025	0.088	0.006	0.0002	0.110	0.002	0.268	0.055	0.129	52		
		21.11.74	5.09	0.032	0.118		0.034	0.099	0.007	0.0005	0.112	0.006	0.145	0.029	0.134	51		
		16.12.74	5.21	0.033	0.110		0.024	0.095	0.007		0.108	0.013	0.108	0.036	0.136	50		
		18.12.74	5.12	0.035	0.128		0.028	0.100	0.005		0.112	0.013	0.112	0.036	0.160	49		
		21.1.75	5.10	0.037	0.120	0.005	0.017	0.096		0.0009	0.110	0.012	0.104	0.034	0.156	50		
		20.2.75	4.70	0.038	0.111		0.023	0.107			0.106	0.017						
		18.3.75	4.82	0.035	0.141		0.019	0.119		0.0054	0.116							
		20.3.75	4.92	0.030	0.122		0.023	0.114		0.0019	0.111							
		12.5.75	5.02				0.036	0.123			0.143				0.142	53		
		15.5.75	4.85	0.010	0.111	<<	0.016	0.111		0.0040	0.125		0	0.027	0.142	49		
		26.5.75	5.09	0.030	0.111	<<	0.016	0.110		0.0043	0.122		0.065	0.040	0.130	48		
		26.5.75	5.14	0.039	0.113		0.031	0.101			0.104		0.179	0.045	0.074	52		
		27.5.75	5.20	0.030	0.130		0.014	0.109		0.0025	0.119		0.026	0.028	0.144	48		
		27.5.75	5.20	0.037	0.108		0.029	0.100			0.105			0.037	0.116	49		

Table 75 (continued)

Group		Date															
B3	60	19.11.74	4.80	0.032	0.102	0.003	0.027	0.074		0.110	0.003	0.294	0.038	0.071	49	0.35	0.47
		21.11.74	5.00	0.035	0.095		0.030	0.078		0.098	0.004	0.108	0.026	0.140	50	0.36	0.42
		16.12.74	5.00	0.033	0.090		0.027	0.071		0.105	0.006	0.080	0.045	0.132	46	0.33	0.39
		18.12.74	5.18	0.037	0.090		0.029	0.080		0.098	0.010	0.089	0.040	0.186	48	0.35	0.50
		21.1.75	5.30	0.036	0.091	0.011	0.041	0.085	0.002	0.103	0.009	0.087	0.035	0.159	49	0.35	0.45
		23.1.75	5.05	0.026	0.092	0.010	0.023	0.095	0.002	0.110	0.016	0.113	0.023	0.168	50	0.38	0.47
		18.2.75	5.00		0.090		0.030	0.100		0.095	0.013	0.113	0.035	0.100	51	0.37	0.36
		20.2.75	4.77	0.031	0.088		0.029	0.095		0.094	0.024	0.076	0.018	0.169	50	0.38	0.39
		18.3.75	4.91	0.029	0.087		0.026	0.109		0.093		0.094	0.023	0.123	51	0.38	0.38
		20.3.75	5.18	0.031	0.092		0.030	0.103					0.027	0.138	51	0.38	
		12.5.75	4.80	0.034	0.092		0.030	0.122		0.112		0.132	0.025	0.141	55	0.43	0.44
		15.5.75	4.90	0.034	0.093		0.032	0.114		0.114		0.104	0.032	0.112	55	0.42	0.36
		26.5.75	4.89	0.033	0.094		0.032	0.114		0.118		0.098	0.026	0.108	54	0.42	0.33
		27.5.75	4.93	0.043	0.087			0.117		0.117			0.025	0.069	54	0.48	
B3+R1	100	19.11.74	4.90	0.046	0.101		0.011	0.056		0.074	0.004	0.238	0.032	0.038	40	0.29	0.34
		21.11.74	5.19	0.039	0.090		0.009	0.059		0.083	0.004	0.177	0.030	0.051	36	0.27	0.31
		16.12.74	4.91	0.036	0.086		0.008	0.060		0.082	0.003	0.201	0.044	0.044	37	0.25	0.38
		18.12.74	5.41	0.037	0.086		0.009	0.050		0.071	0.001	0.147	0.040	0.097	33	0.25	0.36
		21.1.75	5.20	0.030	0.079	0.001	0.023	0.064		0.071	<0.001	0.130	0.043	0.047	34	0.28	0.28
		23.1.75	5.35	0.023	0.076	0.002	0.006	0.053		0.073	0.006	0.100	0.037		31	0.23	0.24
		18.2.75	5.22	0.034	0.086		0.005	0.061		0.100		0.040	0.040	0.064	32	0.25	0.26
		20.2.75	5.02	0.030	0.077		0.005	0.055		0.099		0.071	0.025	0.080	29	0.24	0.26
		18.3.75	5.38	0.029	0.077		0.010	0.062		0.095	0.007	0.124	0.027	0.048	30	0.25	0.25
		20.3.75	5.51	0.028	0.077		0.009	0.059		0.099	0.006	0.098	0.037	0.043	29	0.24	0.23
		12.5.75	5.21	0.031	0.077		0.005	0.065		0.098	0.006	0.136	0.048	0.048	31	0.25	0.29
		15.5.75	5.49	0.030	0.082		0.005	0.060		0.109	0	0.122	0.047	0.046	30	0.24	0.26
		26.5.75	5.37	0.031	0.076		0.004	0.059		0.098	0.002	0.075	0.056	0.056	31	0.23	0.24
		27.5.75	5.71	0.029	0.077					0.113	0		0.055	0.046	29		
		14.7.75	5.33	0.034	0.087		0.007	0.062		0.112	0.005	0.123	0.051	0.045	31	0.26	0.27

Table 76 Chemical composition of the soil waters in the Haarts catchment (PVC-drains)
(<< = << 0.001)

Horizon	Depth (cm)	Date	pH	K^+	Na^+	NH_4^+	Ca^{2+}	Mg^{2+}	Fe	Mn^{2+}	Al^{3+}	H_4SiO_4	NO_3^-	Cl^-	Alk	SO_4^{2-}	EC_{25} µS/cm
						mmoles/l											
A1	10	5.11.73	4.04	0.164	0.044	0.167	0.073	0.026	0.006	0.020	0.022	0.044	<<	0.057		0.321	70
A1	10	10.12.73	4.16	0.118	0.076	0.210	0.058	0.031		0.021	0.026	0.048	<<	0.199		0.196	69
A1	10	8.1.74	4.35	0.105	0.103		0.047	0.033	0.002		0.017	0.015	<<			0.101	50
A1	10	11.2.74	4.21	0.095	0.085			0.023				0.036					63
A1	10	18.3.74	3.92	0.114	0.076		0.118	0.072			0.022	0.009	0.05				116
A1	10	30.5.74	3.70	0.102	0.092	0.057	0.052	0.053				0.070	0.158				118
A1	10	22.10.74	4.50	0.166	0.092	0.119	0.046	0.037	0.003		0.022	0.014	0.125	0.116		0.202	72
A1	10	19.11.74	4.20	0.160	0.109		0.066	0.063	0.001		0.030	0.007	0.152			0.250	95
A1	10	16.12.74	4.48	0.085	0.065	0.060	0.058	0.062				0.008	0.132			0.068	73
A1	10	21.1.75	4.55	0.146	0.077	0.006	0.058	0.032	0.003		0.016	0.069	0.469				43
A1	10	12.5.75	4.50	0.140	0.06		0.062	0.063				0.068				0.190	96
A1	10	26.8.75	4.50	0.100	0.054	0.017	0.045	0.034	0.004	0.010	0.014	0.028	0.033	0.105			44
A1	10	14.10.75	4.52	0.074	0.038	0.003	0.060	0.029	0.001	0.013	0.014	0.007	0.025	0.060			42

Table 77 Chemical composition of atmospheric precipitation samples

Period	pH	K^+	Na^+	Ca^{2+}	Mg^{2+}	NH_4^+	NO_3^-	Cl^-	SO_4^{2-}	$H_4SiO_4^0$	Σ_{cat}	Σ_{an}
				μmoles/l							meq/l	
9/10-5/11 '73	4.50	4	22	13	0.5	4	25	59	43	1.5	0.13	0.17
5/11-6/11	5.50	8	33	11.5	0.2	80.5	12	98	48	0.5	0.15	0.20
6/11-10/12	5.53	1	8	4	0.5	81.5	0	90	11	1.0	0.10	0.11
12/12°	4.65	14	54	18	2.5	127	38	60	55	2.0	0.26	0.21
10/12-7/1 '74	4.48	5	49	30	4	59	16	94	22	0.5	0.21	0.15
7/1-11/2	4.22	10	27	13	3	44	22	78	38	0.5	0.18	0.18
11/2-18/3	4.00	5	16	29	2	91	52	42	47	1.5	0.27	0.19
18/3-22/4*	4.27	20	44	73	13	44	190	92	114	tr	0.33	0.50
22/4-30/5	5.33	23	27	33.5	6	90	29	69	90	tr	0.23	0.28
30/5-1/7	4.92	14	23	20	7	46	21	44	48	1.0	0.15	0.19
1/7-4/7	6.09	8	12	tr	7	90	17	42	45	1.0	0.12	0.15
4/7-7/8*	6.61	67	20	11.5	8	404	6	60	64	0.5	0.52	0.54
8/8	4.70	4	15	9.5	1	22	21	7	6	3.0	0.08	0.08
8/8-22/10*	5.89	15	22	12	7	109	13.5	51	80	1.0	0.19	0.22
23/10	4.41	3	7	6.5	3	15	20	6	22		0.08	0.07
23/10-19/11	5.25	4	38	16	2.5	27	24	32	24		0.12	0.10
19/11-16/12	4.41	2	38	11	5	28	27	55	28		0.14	0.14
16/12-18/12	4.97	2	25	7	6	16	7	80	14		0.08	0.10
18/12-21/1 '75	4.70	10	15	23.5	7	17	17.5	45	46		0.12	0.15
21/1-23/1	5.20	7	43	11.5	2.5	120	14	132	17		0.25	0.18
23/1-18/2	4.70	3	44	12.5	9	tr	9	23	20		0.11	0.10
18/2-20/2	4.50	3.5	22	25	tr	10	28	40	21		0.12	0.11
20/2-18/3	5.18	33	33	37.5	10	65	48.5	15	85		0.20	0.22
18/3-20/3	4.25	10	22	16	tr	13	53	36	18		0.13	0.13
20/3-12/5	4.00	3	16	13.5	tr	54.5	47	25	47		0.20	0.17
12/5-16/7 [(*)]	6.10	19.5	16.5	11	4	133	18.5	14	53		0.20	0.14
16/7-26/8*	6.39	20	16.0	11.5	5	163	19	10	72		0.23	0.17
26/8-15/10 [(*)]	5.80	19.5	43.5	31	6.5	109	91	45	47		0.25	0.23

° snow; [(*)] slightly dirty samples; * dirty samples